TECHNIQUES IN VISIBLE AND
ULTRAVIOLET SPECTROMETRY

VOLUME ONE
STANDARDS IN ABSORPTION SPECTROMETRY

Standards in Absorption Spectrometry

ULTRAVIOLET SPECTROMETRY GROUP

Edited by

C. BURGESS
Glaxo Operations UK Ltd,
Barnard Castle, Co. Durham

and

A. KNOWLES
Department of Biochemistry,
University of Bristol

LONDON NEW YORK
CHAPMAN AND HALL

First published 1981
by Chapman and Hall Ltd.
11 New Fetter Lane, London EC4P 4EE

Published in the U.S.A. by
Chapman and Hall
in association with Methuen Inc.
733 Third Avenue, New York NY 10017

© *1981 UV Spectrometry Group*

Typeset by Tek-Art, Croydon, Surrey
and printed in Great Britain
at the University Press, Cambridge

ISBN 0 412 22470 4

British Library Cataloguing in Publication Data

Standards in absorption spectrometry. — (Techniques in visible and
ultraviolet spectrometry; Vol. I).
1. Absorption spectra
I. Burgess, C   II. Knowles, A   III. Ultraviolet Spectrometry
Group   IV. Series
543'.085   QD96.A2   80-41128

ISBN 0-412-22470-4

Contents

Preface

The Photoelectric Spectrometry Group was formed in July 1948 in Cambridge. The Group was born out of a need for a forum of users to discuss mutual problems and methodology associated with the new era of photoelectric spectrometers heralded by the Beckmann DU absorptimeter. Over the years the aims and objectives of the Group have been broadened to include many aspects of ultraviolet and visible spectrometry. In 1973, the Group renamed itself the UV Spectrometry Group. The techniques of fluorescence, diffuse reflectance and to a lesser extent ORD and CD were included in the Group's interest. In 1979, the UVSG became a registered charity. The present Group membership is some 200 practising spectroscopists, many from the UK plus a small but growing overseas membership.

The Group's active interest in standards and standardization over the last thirty years is readily seen from our Appendix. In August 1977, the committee under the chairmanship of Dr A. J. Everett, Wellcome Research Laboratories, set up three working parties: Cells for UV-Visible Spectrophotometers; Photometric and Wavelength Standards; and the Calibration of Fluorimeters. It was felt that a wealth of information and expertise in the practice of UV spectrometry was available within the Group and that it was appropriate for this to be gathered together in the form of a number of monographs. Initially the intention was that these should be limited to circulation amongst the Group membership. However, the suggestion was made that these monographs would be of interest to other scientists outside our specialist Group.

This monograph is a combination of the first two working party reports and attempts to cover those areas of UV-visible spectrophotometry which are vital to production of accurate and precise data. This is essentially a practising chemists' manual and in no way claims theoretical rigour. Notwithstanding, theoretical aspects have been covered and appropriate references made to more fundamental works in those sections where the reader may require a deeper insight, for example in the chapter on Stray-Light.

We have set out a series of recommended standards for cells and procedures for instrumental standardization. We regard these as much akin to the late J. R. Edisbury's 'Links with Sanity'. The recommendations are a consensus of informed user opinion and do not reflect any commercial interest or bias. An introduction to the topic of UV-visible spectrophotometry has been supplied by Dr Everett in his own highly individualistic style and sets the scene for the later more specialized chapters.

The Cell Working Party discussions were based on British Standard Specification No. 3875 (1965) which led to a unification of cell design and a raising of manufacturing standards. The recommendations of the Deutsches Institut für Normung and the Institut für Standisierung und Dokumentation im Medizinischem Laboratorium contained in DIN 32 635 and DIN 58 963 (Parts 1 and 2) have also been taken into account.

It should be emphasized that this monograph is a product not only of the working parties concerned but also of the 'workshops' which followed the issue of the first drafts. As chairmen of the working parties we are grateful for the interest and expertise displayed at those subsequent meetings in helping to produce the final article. Special mention of individuals is not usual in this type of group effort. However, we wish to acknowledge the extensive assistance of Professor D. Thorburn Burns, Queen's University, Belfast, in providing the Standards Working Party with data and in his critical appraisal of the first draft.

Thanks are due also to the Committee and our present Chairman, Dr M. A. Russell, B.D.H. Chemicals Ltd, who have guided our efforts so ably and to our publishers Chapman and Hall.

The Editors would welcome any comment or criticism concerning this monograph. All proceeds accruing from sales will go to Group funds for the furtherance of UV spectrometry.

February 1980   C. BURGESS

                                                     A. KNOWLES

Membership of the working parties

Cell Working Party

Mr J. H. Barrett, National Physical Laboratory, Teddington, Middlesex

Dr G. H. Beaven, National Institute for Medical Research, Mill Hill, London

Mr S. Beavis, Optiglass Ltd, Walthamstow, London

Mr T. P. Browell, Thermal Syndicate Ltd, Wallsend, Northumberland

Mr D. C. Knowles, Perkin-Elmer Ltd, Beaconsfield, Buckinghamshire

Dr A. Knowles (Chairman), Department of Biochemistry, University of Bristol

Dr A. Moss, Pye Unicam Ltd, Cambridge

Dr T. L. Threlfall, May & Baker Ltd, Dagenham, Essex

Standards Working Party

Dr C. Burgess (Chairman), Glaxo Operations (UK) Ltd, Barnard Castle, County Durham

Dr G. J. Buist, Department of Chemistry, University of Surrey, Guildford

Professor D. Thorburn Burns (*Ex Officio*), Department of Chemistry, The Queen's University, Belfast, Northern Ireland

Dr A. J. Everett (*Ex Officio*), Wellcome Research Laboratories, Beckenham, Kent

Mr D. Irish, Pye Unicam Ltd, Cambridge

Mrs E. Vinter, Wellcome Research Laboratories, Beckenham, Kent

Dr J. G. Vinter, Wellcome Research Laboratories, Beckenham, Kent

1 General considerations on UV–visible spectrometry

1.1 Introduction

Ultraviolet spectrophotometry, as opposed to spectroscopy, has been generally available since about 1943 when it became possible with manual photoelectric spectrophotometers to make reasonably quantitative measurements of the amount of energy absorbed as a function of the wavelength of the incident radiation. Since then a wide range of manual and recording spectrophotometers has become available, but sadly there is no compelling evidence that the reproducibility of measurements between laboratories approaches that from within a given laboratory. Reasonably competent operators seem able to achieve acceptable precision but often only with rather poor accuracy. Many publications have dealt with this problem which of course is the starting point for this Volume. The UV Group has played an active part in the quest for the optimum performance of instruments and a selection of the Group's publications is given in the Appendix.

It is not within our scope to discuss the fundamentals of UV-absorption spectroscopy in terms of the electronic phenomenon. The ramifications of quantum mechanics have little impact upon the finger print on the front surface of a cell. The fundamentals which concern us here are those which bear upon the best use of the available equipment to achieve a spectroscopic measurement. We assume that most readers will have a spectroscopic background and that the following notes will merely serve to jog the memory as well as bring to mind key references [1-4].

1.2 Radiant energy

Three properties of electromagnetic radiation are necessary to specify it. The quantity or intensity is specified in units of energy or power. The quality is defined by the frequency or the vacuum wavelength. Finally, the state of polarization should be specified.

1.2.1 *Wavelength*

In general, the frequency of UV radiation is too high for direct measurement (about 10^{15} Hz) so that experimental measurements must be in terms of wavelength. Frequency is then derived from:

$$\text{frequency} = \frac{c}{\lambda}$$

where c is the velocity of light in vacuum and λ is the wavelength of the radiation.

It is important not to confuse frequency with wavenumber. The latter is the number of wave maxima per unit length, being given by:

$$\text{wavenumber (cm}^{-1}) = \frac{10^7}{\lambda \text{ (nm)}}$$

and although, unlike wavelength it is directly proportional to energy, it has no particular spectroscopic significance.

Visible light is generally considered to extend from 680 nm (14 706 cm^{-1}) to 370 nm (27 027 cm^{-1}) and the near-UV-region from 370 nm to 200 nm (50 000 cm^{-1}). Like the other defining wavelengths, the 200 nm limit is arbitrary in that for old instruments the scattered-light performance rapidly deteriorates with further decreasing wavelength, and oxygen and solvent absorption exacerbate the problem.

Today with the new generation of holographic grating spectrometers a limit of 185 nm might be more realistic. In most instances instrument manufacturers provide spectrophotometers whose precision and accuracy of wavelength read-out are adequate, but this must not dissuade the spectroscopist from simple checks of calibration.

1.2.2 *Intensity*

It cannot be said that the same confidence in wavelength accuracy applies to the measurement of intensity, the second defining property of the radiation. Fortunately for the UV spectroscopist, absolute light intensity measurement rarely arises. It is the attenuation of the light beam which is of more interest to the majority, who are concerned with absorption spectrometry and, here, intensity is loosely equated to absorbance as defined below. If the need does arise, the absolute intensity of the light beam may be expressed in convenient energy units per unit time. The latter aspect will be the subject of detailed consideration in the companion monograph on Fluorescence spectrometry.

1.3   Absorption

1.3.1   *Absorbance*

When a beam of radiation of specific wavelength impinges upon a substance, the energy associated with the beam may be altered by reflection, refraction, absorption and transmission processes.

Most experimental measurements are concerned with elimination of, or corrections for, effects other than absorption. The simplest situation with respect to the intensity of absorption is that in which the system obeys the Lambert-Beer Law. In this case if I_0 is the intensity of a parallel beam of radiation incident normally on a layer of thickness b cm and molar concentration c, the intensity of the emergent beam is:

$$I = I = I_0 \ 10^{-\epsilon c b}$$

where ϵ, the molar absorptivity (litre mole^{-1} cm^{-1}), is independent of c but is a function of wavelength, temperature and solvent.* Of course this implies that each layer, or indeed each molecule, of the absorbing substance absorbs a constant fraction of the incident radiation. The above equation can be expressed in the form:

$$\log_{10} \left(\frac{I_0}{I} \right) = \epsilon c b$$

or

$$A = \epsilon c b$$

where A is the absorbance of the sample in the beam. The ratio of the light intensity transmitted by the sample to the light intensity incident on the sample is the transmittance T:

$$T = I/I_0 \qquad \text{and} \qquad A = -\log_{10} T$$

In this Volume, transmittance will be expressed as a percentage, i.e. $T = 100 \, I/I_0$ per cent.

---

*The nomenclature used throughout is based on the recommendations of the American Chemical Society [5], the American Society for Testing and Materials [6], and the British Standards Institution [7].

Absorbance is more simply related to concentration and absorptivity than are I, I_0 or T. Strictly, absorbance is only applicable to solutions, the more general term 'optical density' applying to solids and homogeneous liquids as well. However, absorbance will be taken to be synonymous with optical density.

The attenuation of a beam of radiation in passing through a sample is due in part to absorption within the sample and in part to reflection and scatter at the external surfaces. The transmission of the material itself, without the external losses, will be termed the 'internal transmission', and is thus defined as that percentage of the radiant flux leaving the entry surface which eventually reaches the exit surface.

1.3.2 *Sources of absorbance error*

It is convenient to consider two categories of absorbance error. The first originates with the spectrophotometer and the second directly or indirectly with its use. In practice this dichotomy is not so clearly defined.

(a) *Spectrophotometer limitations*

At the outset it is desirable to distinguish between the working definition of transmittance or absorbance and the true transmittance or absorbance as outlined by Jones and Sandorfy [2]. Using their terminology, for parallel radiation of intensity I_i falling normally on a cell containing a solvent and a solute:

I_r = reflection losses at cell interfaces
I_s = scattering losses at cell surfaces and from the solution
I_b = absorption losses by the solvent
I_a = absorption by the solute.

The true transmittance of the solute is:

$$T = \frac{I_i - (I_a + I_b + I_r + I_s)}{I_i - (I_b + I_r + I_s)}$$

On the other hand the working definition of transmittance, T', generally using a double-beam technique, is:

$$T' = \frac{I}{I_0} = \frac{I_i - (I_a + I_b + I_r + I_s)}{I_i - (I'_b + I'_r + I'_s)}$$

It follows that T and T' are only identical when:

$$I_b + I_r + I_s = I'_b + I'_r + I'_s$$

Deviations from this condition are most likely to occur for a sample with low molar absorptivity and high molecular weight.

(b) *Reflection losses*
If sample and reference cells are made to a sufficiently high specification the outer face reflections will cancel. So also will the reflections at the liquid-to-fused-silica interfaces if, as is usual in UV spectrophotometry, the solute concentration is very low. To put the matter into perspective, the loss from internal reflections in a fused silica cell filled with water is only about 0.4 per cent of the incident light energy at 589.3 nm [8].

Even on passing through an absorption band where the solution refractive index and hence the reflectance loss (see Section 1.4.2) is rapidly changing, the effect being on the solute absorbance measurement is exceedingly small, being of the order of 0.001 per cent. Measurable effects arising from refractive index imbalance between reference and sample do arise, but they are essentially of an instrumental nature and may be responsible for some of the difficulties associated with the use of potassium nitrate solutions as absorbance standards where the concentrations are as high as 0.15 M.

(c) *Solvent absorption*
Usually in UV spectrometry the mole fraction of the solute is so low that $I_b = I'_b$, i.e. the numbers of absorbing solvent molecules in each beam are almost identical.

(d) *Scattering losses*
Small non-conducting particles will, when present as a cloudy sample, exhibit Tyndall scattering whose intensity is proportional to the fourth power of the frequency. This can give rise to very serious problems which lead to apparent deviations from the Beer-Lambert Law, particularly at short wavelengths. Good working practice will reduce the gratuitous introduction of scattering errors.

1.4   User limitations

The following factors are ones which should be considered when attempting to obtain the greatest precision and accuracy from a spectrophotometer. In some instances the instrument design will dictate procedure; in others the user can have a marked influence on the quality of the result.

1.4.1   *Gravimetric and volumetric accuracy*

This is not the place to deal with these factors *in extenso*. Suffice to note that in most instances weighing errors certainly ought not to exceed 0.1 per cent and volumetric errors should be little more, unless very small volumes are to be handled in which case the solvent should be weighed and corrections applied for solvent density. Solvation, particularly hydration, is a frequent source of error associated with the measurement of molar absorptivity. It must also be borne in mind that appreciable temperature changes, besides affecting volumetric equipment, will frequently lead to actual changes in the molar absorptivity of the sample. For example it has been suggested that the latter is partly responsible for the apparent variability of the molar absorptivity of aqueous potassium nitrate. Particulate matter can be removed by using a proprietary membrane filter on a plastic hypodermic syringe. The syringe should be washed before use, as it may be coated with a lubricant. Glass hypodermic syringes may introduce fine particles of glass and should never be used for transferring solutions. Small bubbles of air adhering to the window surfaces are a source of error exacerbated by greasy surfaces but alleviated by solvent degassing and careful manipulation of the solutions.

1.4.2   *Cell handling*

An obvious first requirement of any photometric measurement of solutions is that the effect of the container should be measurable or compensated. Ideally, the sample and reference cells should be optically identical. Apart from the identity of the window geometries and the consistent orientation of the cells with respect to the light beams, it is an elementary requirement that they be clean. Recommended cleaning procedures are given in Chapter 8.

The ratio of the reflected light intensity I_r to the incident light intensity I on a surface is governed by the Fresnel relationship:

$$\frac{I_r}{I} = \left(\frac{n_1 - n_2}{n_1 + n_2} \right)^2$$

where n_1 and n_2 are the refractive indexes of the two media. Other than at short wavelengths the transmittance of empty synthetic fused silica will be governed by reflectance losses. For example, the theoretical transmittance of an empty cell at the wavelength of the sodium D line (589.3 nm) is 0.933 (A = 0.0301) whereas on filling with water† the transmission rises to 0.996 (A = 0.018) because the internal reflection losses are decreased as a consequence of the nearer matching of the refractive index of water to that of fused silica. However in practice an empty cell need never be used in the reference beam of a spectrophotometer. Ideally, cells should only be handled with tongs or hands covered by clean cotton gloves.

Accuracy is required in setting the cell in the beam [3]: For an absorbing medium of refractive index n and for an angle of a radians from the normal to the cell with respect to the incident beam, the fractional error in path length δ, is given by:

$$\delta = \left(\frac{0.0123a}{n} \right)^2$$

For many advanced instruments an absorbance of 2 can be read to at least 0.001 A (δ = 0.0005). Even simpler instruments can approach this precision when properly used in the difference mode. An alignment error of just over 3° would introduce an error of this magnitude. To put the matter in perspective there would have to be a side play of 1 mm in a 1 cm cell-holder to introduce this error. Nonetheless the hazard of using cells with non-standard external dimensions should be recognized. It is also wise to check that the mounting of short cells is such as to give adequate reproducibility and that micro cells and long cells do not attenuate either by vignetting the beam at the front window or by the light beam grazing the cell walls.

In most spectrophotometers the cells are mounted symmetrically with respect to the focus of the beams and pathlength. Errors arising from beam divergency are usually negligible.

For the most precise work with dilute solutions it is probably best to use only one cell and leave it in position for both solvent and solute readings. A hypodermic syringe with a plaster catheter tube is then used for filling, emptying and washing the cell. The solvent spectrum readings are subsequently subtracted from those of the solution. Spectrophotometers interfaced to computers now make this approach quite attractive to the user if not to the cell manufacturer. The cell 'blank' can be stored so that the experiment time remains the same

---

†Water must be stored in fused silica vessels or, failing that, polyethylene which contains neither antioxidant nor plasticizer.

as the conventional method with cells in both beams. Equipment is available to allow the whole process to be automatically controlled. For those common solvents which are transparent over a large wavelength range (octane, water, ethanol) there is little point in insisting on very close matching of the cell lengths. Good window parallelism, composition, and surface finish are far more important.

Even with perfectly matched cells some double-beam spectrophotometers are sufficiently sensitive to small refractive index differences between the contents of the two cells to give rise to absorbance errors. This instrumental effect occurs because the pathlength between cell and detector is often long, to limit the effects of scattering and fluorescence, and provides an optical lever which moves the light image on the photocathode according to the refractive index in the cell. Unless the photocathode response is absolutely uniform over its surface, the beam movement will appear as an absorbance change. These matters are further discussed in Chapters 3 and 6.

1.4.3 *Choice of absorbance*

Spectrophotometers do not provide constant precision throughout their absorbance range. Therefore to achieve the best analytical quantitative performance the combination of solution concentration and cell length should be adjusted to the most precise region of the instrument absorbance scale. To a great degree this region is governed by the noise characteristics of the detector which, for most modern recording instruments, is a photomultiplier. For a photomultiplier detector the noise, N, is related to the incident intensity in the form:

$$N = K\sqrt{(\text{intensity})}$$

where K is a proportionality constant. By differentiating the Beer-Lambert relationship and recognizing that noise, the error in measuring intensity, obeys the above relationship, the error, P, in measuring concentration can be derived in the form:

$$P = K \left[\frac{\sqrt{(I_0)}\ A}{10^{A/2}} \right]^{-1} \tag{1.1}$$

This function has a broad minimum at $A = 0.869$ (13.5 per cent transmission). In practice it is probably sufficient to use an absorbance between 1 and 2. However if stray-light errors are likely, for example when using a strongly absorbing solvent, or the spectrophotometer is being used at the extremes of the range of the grating, prism, detector or source, the higher absorbances should be avoided.

The choice of optimum conditions is not simple. There are many interdependent variables such as cell pathlength, concentration, spectral slitwidth, scattered radiation, scale read-out discrimination. A number of publications deal with the matter in some detail [1, 2].

1.4.4 *Choice of slitwidth*

When we set a monochromator to a nominal wavelength, an approximately triangular intensity distribution of wavelengths emerges. That wavelength range which never contains less than half of the peak emergent light energy is termed the effective spectral slitwidth (ESW). The spectral region isolated, i.e. the width of the image of the exit slit along the wavelength scale, is termed the spectral slitwidth (SSW). The natural bandwidth (NBW) of a band measured in the spectrum of a compound is the width at half the height, measured at infinite resolution. The ratio ESW/NBW decides how closely the spectrophotometer measurement approaches the true height. When the effective slitwidth is about 12 per cent of the new NBW, the spectrometer will read about 99 per cent of the true height. A number of workers have computed the effect of finite spectral slitwidth on band shape and intensity. The most recent by Torkington [9] provides data for both Gaussian and Lorentzian band profiles as well as the following analytical expression relating ESW/NBW to the fraction of the true height for a Lorentzian profile:

$$F = \frac{2}{q} \tan^{-1} q - \frac{1}{q^2} \ln (1 + q^2)$$

where F is the fraction of the true height at the peak maximum for ESW/NBW = $q/2$

Although UV absorption bands are usually more Gaussian than Lorentzian, this equation holds to within 0.2 per cent up to ESW/NBW = 0.25 and even at ESW/NBW = 0.5 the error is only 1.6 per cent. Numerical values of F as a function of the effective slitwidth are given in Table 1.1.

It is worth noting that the fall in intensity of an absorption peak with SSW is relatively modest compared with the considerable

Table 1.1: *The reduction in apparent height of Gaussian and Lorentzian peaks with increasing slitwidth, from Torkington [9].*

$\dfrac{ESW}{NBW}$	F	
	Gaussian	Lorentzian
0.100	Not calculated	0.9934
0.125	0.9897	0.9899
0.250	0.9603	0.9620
0.500	0.8615	0.8776
1.00	0.6367	0.7048
1.50	0.5203	0.5768

reduction in noise that ensues. I_0 in Equation 1.1 is proportional to the square of the slitwidth and therefore:

$$ P = K' \left[ESW \; \frac{10^A}{10^{A/2}} \right]^{-1} $$

If we are in the habit of maintaining the effective slitwidth at about 0.5 nm and now increase it to be about equal to the natural band-width, say, 25 nm, the fall in height of the peak will be about 64 per cent compared with a fifty-fold gain in sensitivity. This is an extreme example intended to draw attention to the general phenomenon and would only apply if noise is the sole factor which is limiting the sensitivity. It is worth noting that, for an isolated absorption band, the relationship between peak height and concentration will be linear regardless of the value of the SSW, provided that incident radiation all lies within the wavelength boundaries of the absorption band (zero stray-light). The same is true for overlapping bands of the same component, but of course the observed proportionality constant will not be the true molar absorptivity. This does not apply to band overlaps from mixtures of components where the effect of high ratios of ESW/NBW will lead to apparent deviations from the Beer-Lambert Law.

When recording spectra with the intention of obtaining the most accurate value of the molar absorptivity at the peak maximum, it takes little extra effort to measure the absorbance as a function of slitwidth. A typical relationship is shown in Fig. 1.1 where the centre of the plateau is obviously a desirable point at which to make

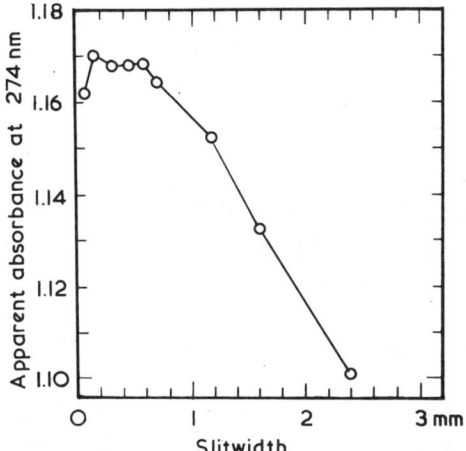

Fig. 1.1   *The effect of increasing slitwidth upon the apparent height of a narrow absorption band. The measurements were made on a Beckman Acta CV with the wavelength set at the maximum of the band [8].*

measurements. Excessive zeal in closing slits will lead to diffraction errors which will cause a fall from the plateau readings.

1.4.5   *Stray-light*

The wavelength range of a spectrophotometer is largely determined by the energy distribution of the source in relation to the transmission characteristics of the monochromator and the spectral response of the detector. When the spectrophotometer is operated under conditions where any one of these is approaching its wavelength limit, stray-light errors may arise. The fractional effective stray-light of a spectrophotometer is the relative proportion of the detector signal which arises from light scattered within the monochromator (other than that of the nominal wavelength).

The stray-light fraction y sets a limit to the absorbance range, for clearly the spectrophotometer cannot respond to absorbances higher than $-\log y$. Nearly always stray-light leads to low absorbance values (negative deviations from Beer's Law) and, in situations where the stray-light fraction is changing rapidly with wavelength, to errors in the wavelengths and shapes of absorption bands. The Beer's Law deviations increase with increasing absorbance. Ultimately, when the sample absorbance is very high, any light transmitted must originate from unwanted radiation and the measured transmittance will approximate to the stray-light fraction:

$$y = \frac{I_s}{I_0 + I_s}$$

where I_s is the intensity of stray-light and I_0 is the intensity of wanted light. I_s sets the dynamic range of the instrument. At short wavelengths (220 nm and below) the incident energy I_0 decreases continuously with wavelength and y rises. At wavelengths below 200 nm I_0 will be reduced even further as a consequence of oxygen absorption and, in this region, nitrogen purging of the optical system is desirable.

An important non-instrumental factor which is of great relevance at short wavelengths is the use of absorbing solvents. They lead to a reduction in the proportion of wanted radiation:

$$A_x = \log_{10}\left[\frac{(1-y)T_s}{T_x' y + (1-y)T_s T_x' - y}\right]$$

where A_x = true absorbance of a solute 'x' in a solvent; T_x' = apparent transmittance of solute 'x' measured with solvent reference; T_s = true solvent transmittance, and y = stray-light fraction. When the solvent is transparent ($T_s = 1$), the equation reduces to:

$$A_x = \log_{10}\left[\frac{1-y}{T_x' - y}\right]$$

It is evident that, even for a stray-light fraction of 0.0015 (0.15 per cent), a solvent absorbance of 1 would raise the error in the measurement of an absorbance of 0.8 from 0.4 per cent to 4 per cent. In a poorly maintained optical system a stray-light fraction of at least 0.01 (1 per cent) may be encountered at wavelengths close to 200 nm, in which case the corresponding absorbance errors become 2.8 per cent and −21.5 per cent respectively. Ultimately the situation can become so bad that if, otpimistically, the monochromator is set at 185 nm there may be so little of this radiation present that the absorbance observations derive from some indeterminate nearby longer wavelength. In these circumstances one would be better off setting the monochromator to a longer wavelength, even though the molar absorptivity is likely to be lower. To resort to a double monochromator would be an expensive solution to the problem. A cheaper alternative is to avoid working too close to the transmission cut-off of the solvents, as shown in Table 1.2, and to note that stray-light errors are reduced as the absorbance of the solution falls.

Table 1.2: *The effective cut-off points for some common solvents. Data from the* UV Atlas of Organic Compounds [10].

Solvent	10 mm pathlength		1 mm pathlength	
	v (cm^{-1})	λ (nm)	v (cm^{-1})	λ (nm)
n-Hexane	50 200	199.2	52 500	190.5
n-Heptane	50 000	200.0	53 000	188.7
Iso-octane	49 500	202.0	53 000	188.7
Diethyl ether	48 700	205.3	50 200	199.2
Ethanol	48 200	207.5	51 000	196.1
Iso-propanol	47 800	209.2	–	
Methanol	47 500	210.5	51 000	196.1
Cyclohexane	47 200	211.9	51 000	196.1
Acetonitrile	46 900	213.2	49 800	200.8
Dioxan	46 300	216.0	48 000	208.3
Dichloromethane	43 000	232.6	–	
Tetrahydrofuran	42 000	238.1	–	
Chloroform	40 500	246.9	–	
Carbon tetrachloride	38 900	257.1	–	
Dimethyl sulphoxide	37 000	270.3	–	
Dimethyl formamide	36 900	271.0	–	
Benzene	35 700	280.1	–	
Pyridine	32 700	305.8	–	
Acetone	30 200	331.1	–	

v and λ are the values of wavenumber and wavelength at which the transmittance falls to 25 per cent (A = 0.602) in the given pathlength, measured using water as the reference.

For example, the errors in the previous example would have been reduced to 1.1 per cent and 10 per cent respectively if the solution concentration or optical pathlength had been reduced by a factor of 10 to give a nominal absorbance of 0.08. The stray-light problem is discussed in detail in Chapter 6.

1.4.6 *Solvents*

A wide range of solvents is available for UV spectroscopy. Since the transmission of most of these falls at the shortwave end of the range, stray-light errors will become significant if the solvents are used incautiously in this region. As indicated in Section 1.4.5, the effect of a solvent absorbance of 1 in the reference cell will be to increase the effect of stray-light approximately ten-fold. A particularly useful set of solvent spectra is provided in the *UV Atlas of Organic Compunds* [10] from which the cut-off wavelengths in Table 1.2 have been derived. Abandoning the ubiquitous 10 mm cell for shorter cells is strongly recommended, though care in cleaning is essential and may be more exacting. The higher solute concentrations rarely give rise to solubility problems but, in regions of high solvent absorption, the usual care concerning the difference in solvent content between the sample and reference cells must be recognized. The single-cell technique with subsequent solvent subtraction is probably the best.

When working near to the cut-off of ethers and hydrocarbons it should be noted that dissolved oxygen greatly reduces their transmission. Degassing with nitrogen is a simple and undemanding precaution.

1.4.7   *Instrument purging*

It is my experience that most spectrophotometers should be purged with clean filtered nitrogen from the time of installation, even if work at short wavelengths is likely to be infrequent. An old instrument should not be purged unless it has been thoroughly cleaned, otherwise the nitrogen flow will merely spread the dust over the optical surfaces. At least the lamp house should be purged to reduce derioration of the source optics.

Unfortunately most lamp houses have cooling vents which circulate air and dust over the mirrors. Cleanliness of these mirrors external to the monochromator is a vital factor in maintaining the stray-light performance of many instruments. Nitrogen purging also reduces the interference by the oxygen bands in the region below 200 nm.

1.4.8.   *Monitoring procedures*

Modern spectrophotometers are long-suffering devices which will continue to give approximate readings long after the uncaring user deserves them. Regular instrument monitoring employing absorbance, stray-light and wavelength checks are relatively undemanding procedures which will expose the false security of blind acceptance. The interval between tests should not be more than a week and, for critical work, less. Most spectrophotometer manufacturers now provide simple check procedures. A useful addition is the installation of a clock activated by the deuterium arc circuit. Sputtering and evaporation on to the inner surface of the silica envelope surrounding the arc leads to a steady reduction in the light flux with time. For many arcs the emission below 220 nm can become seriously low after 100 hours of use.

It is strange that many will find the capital for the purchase of a spectrophotometer and accessories but draw the line at the relatively small cost of certified filters from the National Physical Laboratory (NPL) of the National Bureau of Standards (NBS) for absorbance calibration. For those Scrooges, solution tests with potassium dichromate are better than nothing at all.

1.5   Good spectroscopic practice

Ensure that:

(a) the solution concentration is free from weighing, volumetric and temperature errors;

(b) the compound is completely dissolved: ultrasonic treatment as a routine is provident;

(c) the solution is not turbid - filter if necessary - and that there are no bubbles on the cell windows;

(d) adsorption on cell walls is not occurring;

(e) the cells are clean and oriented consistently in the light beam;

(f) the reference solvent is subjected to *exactly* the same procedures as the solution;

(g) the effective slitwidth is correct for the expected natural bandwidth if absorbance accuracy is important;

(h) important regions of the spectrum are measured with the sample absorbance lying between $0.8A$ and $1.5A$. Adjust the cell length rather than the concentration;

(i) stray-light is not responsible for negative deviations from the Beer-Lambert Law at high absorbance, particularly if the solvent absorbs significantly;

(j) regular tests of absorbance and wavelength accuracy are carried out, and check that stray-light is within specification;

(k) the manufacturer's recommendations concerning scanning speed and time constant are observed;

(l) the environment of the instrument is clean and free from external interference. Particular attention should be paid to electrical interference, thermal variations and sunlight.

References

1 Bauman, R.P. (1962), *Absorption Spectroscopy,* John Wiley & Sons, New York and London.
2 West, W. (Ed.) (1956), *Technique of Organic Chemistry, Vol. IX: Chemical Applications of Spectroscopy,* Wiley-Interscience, New York and London.
3 Lothian, G.F. (1958), *Absorption Spectrophotometry,* Hilger and Watts, London.
4 Richardson, J. and Peterson, R.V. (Eds) (1974), *Systematic Materials Analysis, Vol. II,* Academic Press, New York and London.
5 Spectrometry nomenclature (1980), *Anal. Chem.,* **52**, 221.
6 *Tentative definitions of terms and symbols relating to molecular spectroscopy* (1968), American Society for Testing and Materials, publ. E131-68.
7 *Glossary of Electrotechnical, Power, Telecommunication, Electronics, Lighting and Colour Terms* (1971), B.S. 4727, Pt 4: *Terms particular to lighting and colour;* Group 01: *Radiation and Photometry,* British Standards Institution, London.
8 Everett A.J., unpublished.
9 Torkington, P. (1980), *Appl. Spectr.,* **34**, 189.
10 Perkampus, H., Sandeman, I. and Timmons, C.J. (Eds) (1971), *UV Atlas of Organic Compounds,* Butterworths, London; Verlag Chemie, Weinheim.

2 Cell design and construction

2.1   Introduction

A cell is a container for liquid samples especially made for the measurement of absorption or fluorescence in the wavelength range 180-1000 nm, that is, the ultraviolet, visible and part of the near-infrared regions of the spectrum. This chapter deals with the types of cell commonly used in currently available commercial instruments, and does not attempt to review all possible types of window material or cell design. Typical cells are shown in Figs 2.1, 2.2 and 2.3.

In specifying a cell it is necessary to detail the overall design, the material of construction, the method of assembly, the transmission characteristics of the completed cell, the pathlength and the dimensional tolerance. While attempts were made in BS 3875 to specify also the quality and surface finish of the windows, this Working Party considers that these properties cannot be quantified, but since they have a direct bearing on the performance of the cell, specification of the latter is sufficient. Other aspects of the quality of manufacture, e.g. freedom from pin-holes, mechanical strength etc., are even more difficult to specify, and it becomes incumbent on the user to satisfy himself that the cells are of sufficient quality for his purposes.

2.1.1   Definitions

(a)   *Type of cell*
The type of cell refers to its design as far as this relates to the purpose for which it is intended to be used. Various cell types are described in Section 2.1.2.

(b)   *Grade of cell*
The grade of cell refers to the quality of the cell are far as the dimensional tolerances are concerned. This will, in turn, have a bearing on

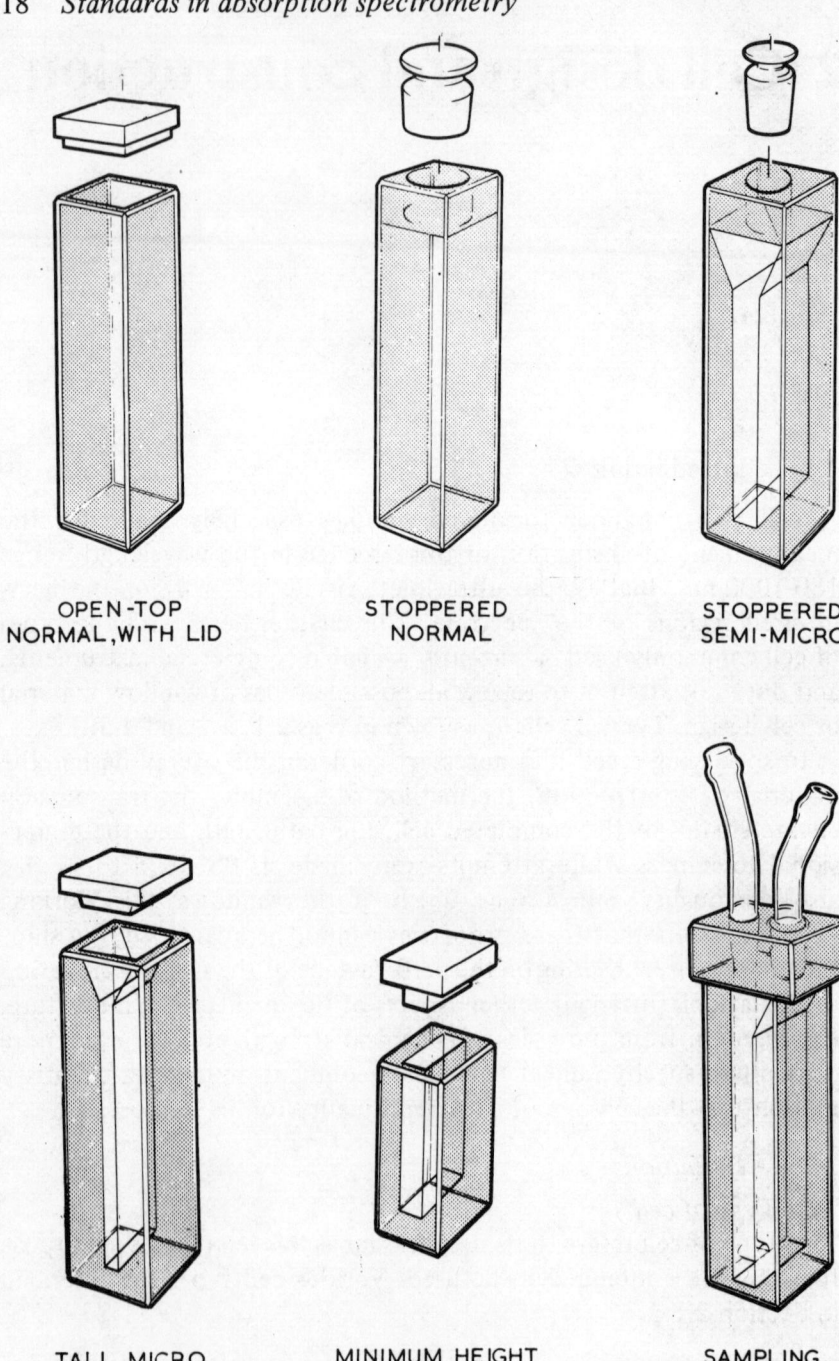

OPEN-TOP
NORMAL,WITH LID

STOPPERED
NORMAL

STOPPERED
SEMI-MICRO

TALL MICRO

MINIMUM HEIGHT
MICRO

SAMPLING

Fig. 2.1   *Representative designs of rectangular cells.*

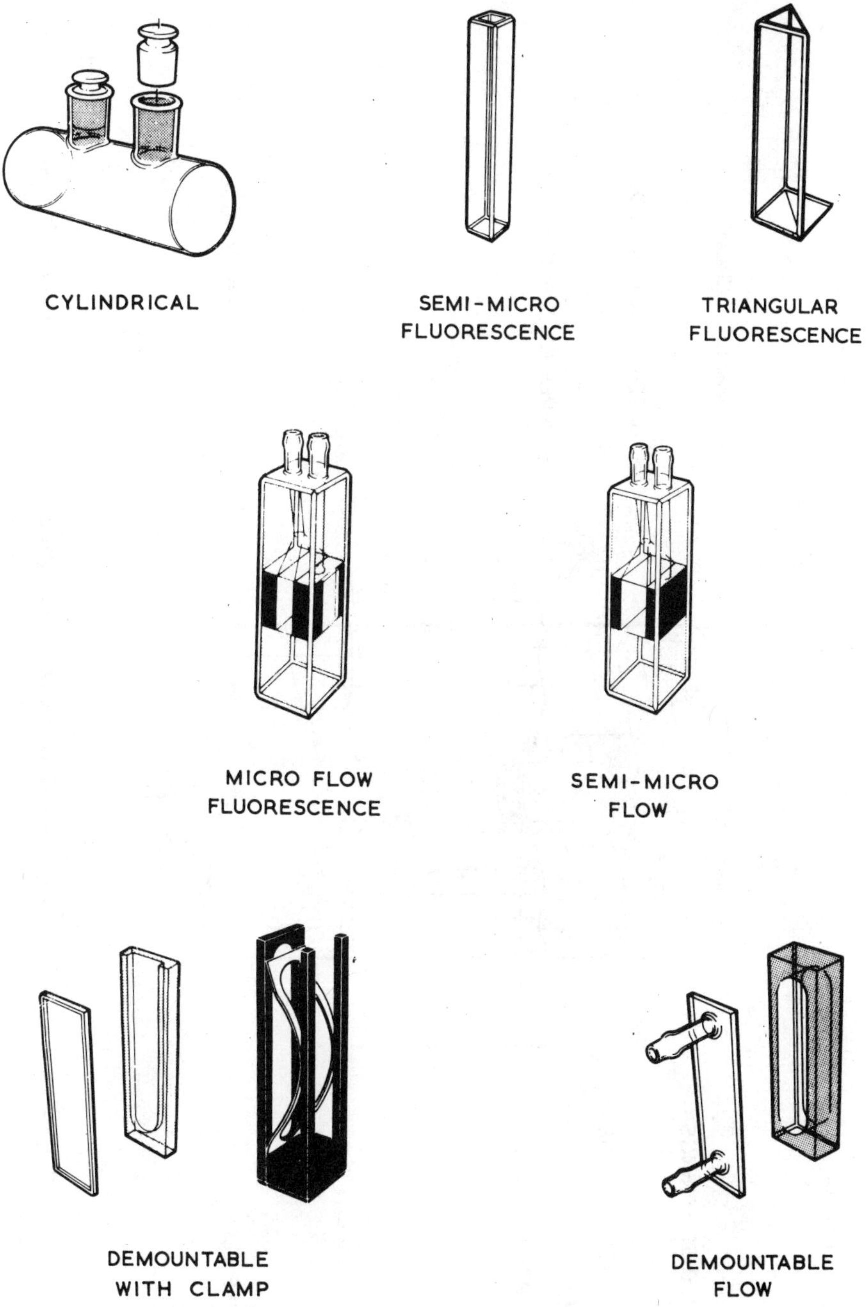

CYLINDRICAL

SEMI-MICRO
FLUORESCENCE

TRIANGULAR
FLUORESCENCE

MICRO FLOW
FLUORESCENCE

SEMI-MICRO
FLOW

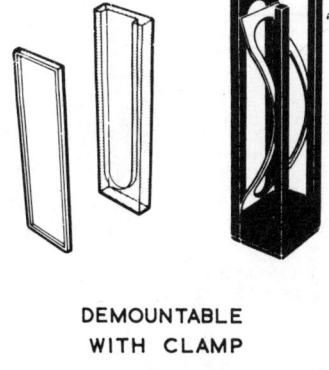

DEMOUNTABLE
WITH CLAMP

DEMOUNTABLE
FLOW

Fig. 2.2   *Representative designs of other types of cell.*

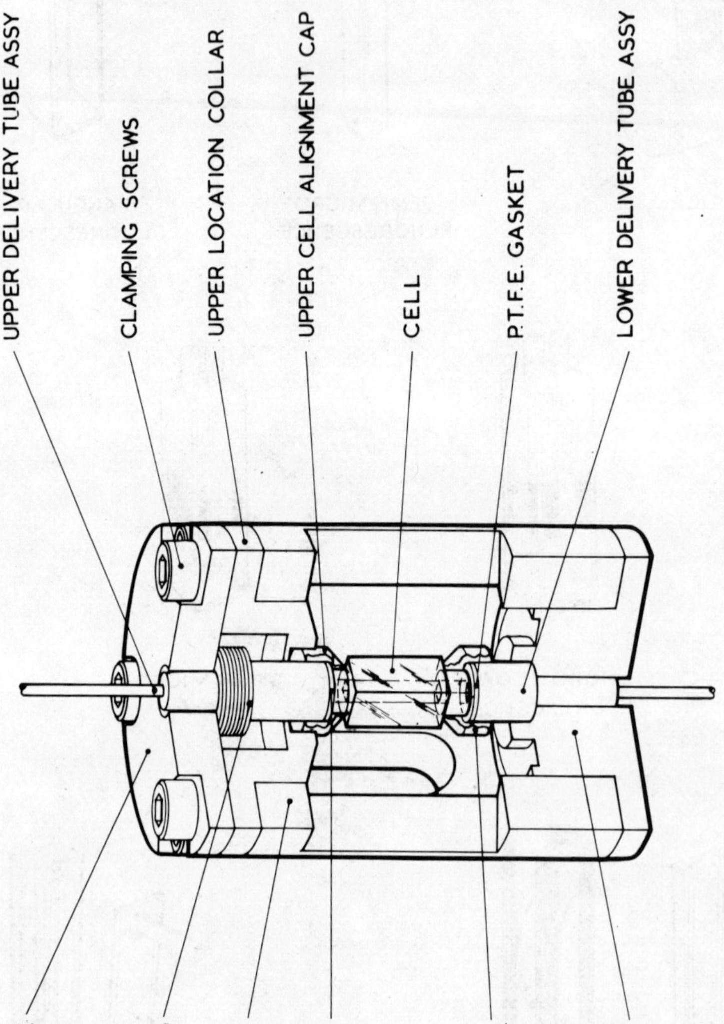

UPPER DELIVERY TUBE ASSY

CLAMPING SCREWS

UPPER LOCATION COLLAR

UPPER CELL ALIGNMENT CAP

CELL

P.T.F.E. GASKET

LOWER DELIVERY TUBE ASSY

TOP PLATE

DISC SPRINGS

MOUNT TUBE

P.T.F.E. GASKET

LOWER CELL ALIGNMENT CAP.

LOWER LOCATION COLLAR

Fig. 2.3   *A micro flow cell intended for fluorescence measurements under pressure in an HPLC system. (Diagram by courtesy of Perkin-Elmer Ltd)*

the use of a cell, its material and method of construction, and its price. Three grades of cell are described in Section 2.1.3.

(c) *Cell windows*
A cell must have two optically clear windows or end plates arranged parallel to one another and perpendicular to the measuring beam. A cell for fluorescence measurement generally has additional windows to transmit the emitted radiation.

(d) *Cell walls*
The structural parts of the cell that are not required to transmit radiation.

(e) *Working area*
An area in the windows of a cell which is intended to encompass the measuring beam and which conforms to the tolerances applicable to that grade of cell.

(f) *Measuring beam*
The instruments referred to in this Volume are designed to pass an approximately collimated beam of light through the working area of the cell when this is placed in the cell-holder of the instrument. In practice, the beam shape is often an enlarged image of the mono-chromator exit slit, that is, an upright rectangle. The beam may well converge or diverge by a few degrees during its passage through the cell.

(g) *Fluorescence*
Some substances absorb UV or visible radiation and re-emit it at a longer wavelength. Such emission will be termed 'fluorescence' in this Volume. It is emitted in all directions, but in most commercial instruments is measured only in a direction perpendicular to the incident beam.

(h) *Angular deviation*
If the two faces of a window are not parallel, or the windows are not parallel to each other, the measuring beam will be deflected in passing through the cell. This deflection may be in the horizontal or vertical plane, but in most instruments an angular deviation in the horizontal plane will have a greater effect on the accuracy of measurement. The deviation angle varies with the refractive index

of the liquid in the cell, and this must be specified when quoting values for the angle.

(i) *Dispersion*
Optical defects in the window may cause distortion of the beam as it passes through the cell. Different kinds of refractive faults may give rise to this distortion, which will be generally termed dispersion.

(j) *Cell dimensions*
These are named in Figs 2.4 and 2.5.

2.1.2 *Cell types*
The cells dealt with in this Volume are classified on the lines of the following features:

(a) *Static, sampling or flow*
'Static cells' are simple containers which are filled and emptied manually, and may or may not be removed from the instrument for refilling. 'Sampling cells' are fitted with tubes so that they can be filled and emptied *in situ* by pressure or vacuum. Usually they must be emptied as completely as possible before refilling, since they are not designed so that a new sample fully displaces the previous one. 'Flow cells' are intended for continuous flow operation and are designed so that each sample completely displaces the preceeding one. This requires that the cross-section of the chamber is a minimum, and that there are minimal dead spaces in the path of the flow. They may be used with a continuously varying sample, as in a chromatography column monitor, or with discrete samples separated by air bubbles, as in an auto-analyser. Examples of these three types are shown in Figs 2.1, 2.2 and 2.3.

(b) *Absorption or fluorescence*
An absorption cell is used to measure the transmittance of a liquid, the measuring beam passing straight through it. A fluorescence cell has additional windows, at the bottom or sides, to permit observation at right-angles to the incident beam.

(c) *Rectangular or cylindrical*
A rectangular cell has flat rectangular walls and so the windows and the working area are rectangular. The body of a cylindrical cell is formed from a length of tubing, with circular windows joined

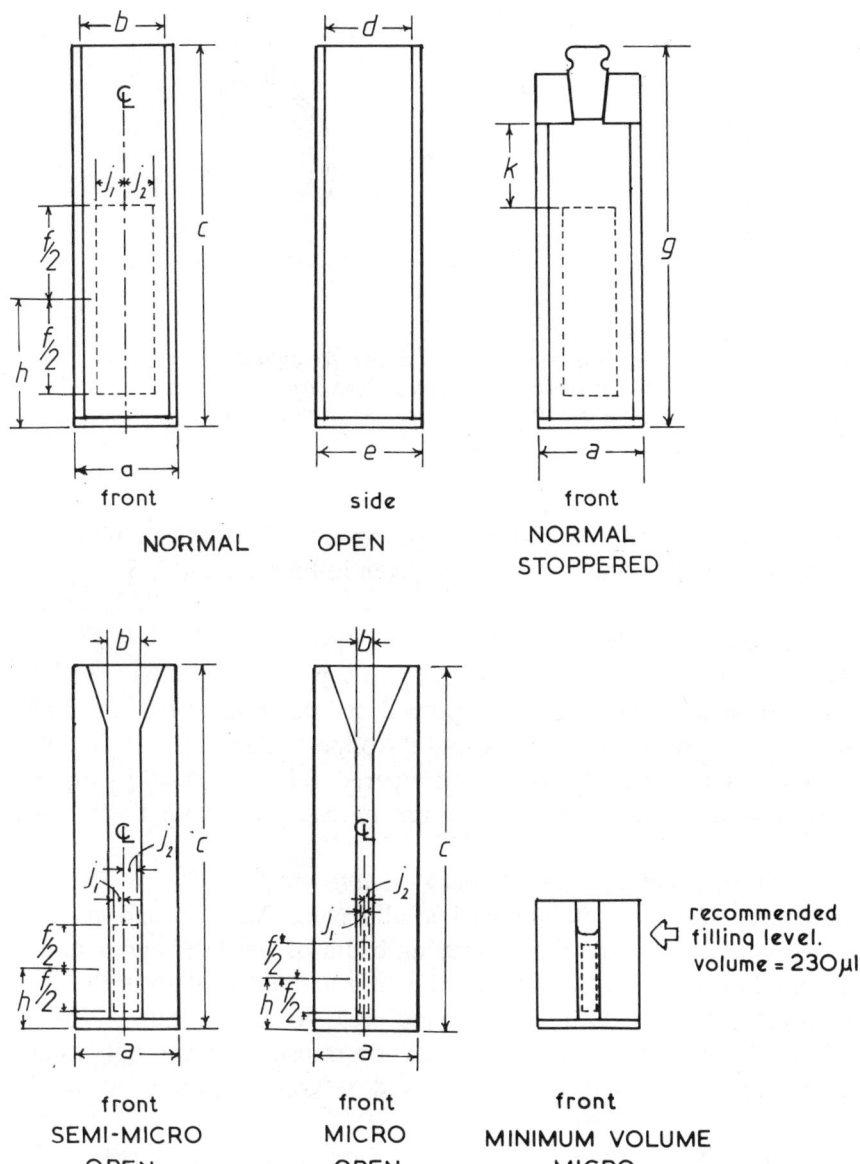

Fig. 2.4 *Dimensions and recommended designs for rectangular cells. The broken lines show the working area. (a) external width; (b) internal width; (c) height of cell body; (d) pathlength; (e) external length; (f) height of working area; (g) overall height including stopper; (h) height of centre of working area above base (normally 15 mm); (j) distance of sides of working area from centre-line of cell; (k) distance between top of working area and top of chamber.*

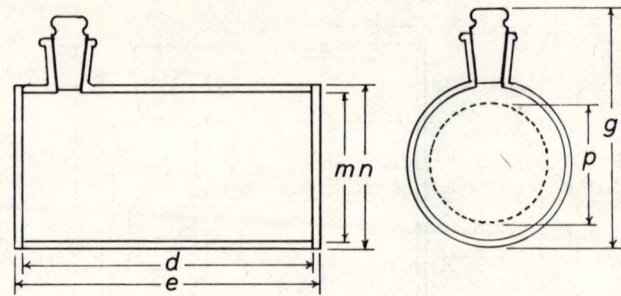

Fig. 2.5   *Dimensions and recommended design for cylindrical cells. The broken line shows the working area. (d) pathlength; (e) external length; (g) overall height including stopper; (m) internal diameter; (n) external diameter; (p) diameter of working area.*

on, and therefore has a circular working area. The designs of typical rectangular and cylindrical cells are given in Figs 2.4 and 2.5.

(d)   *Open or stoppered*
An open cell has a large opening at the top which may be covered - but not sealed - by a lid. A stoppered cell has one or two openings that may be sealed with liquid-tight stoppers. The latter are usually of a standard tapered pattern. Stoppered cells are usually supplied with stoppers, but open cells are not necessarily supplied with lids.

(e)   *Normal, semi-micro and micro rectangular cells*
The best results are obtained with a cell having a large enough working area to accommodate the measuring beam of the instrument under all conditions of its operation. The internal width of the cell must be greater than that of the working area, to ensure that the beam cannot be reflected from the walls, or transmitted through them. Such a cell is classed as *'Normal'* and a typical example is shown in Fig. 2.4.

When only a limited volume of solution is available, a cell can be used in which the volume is reduced by making the chamber narrower. This reduces the width of the working area. The cell is not usually filled to the same depth as the Normal cell, so the height of the working area and its distance above the base are also reduced. Such a cell is termed *'Semi-micro'* and an example is shown in Fig. 2.4. They can be used in most spectrophotometers under normal conditions of operation without special precautions, but the operator should check that the beam does not interact with the walls during the measurement.

A greater reduction in volume can be achieved by further reducing the working area. This is termed a *'Micro cell'* and an example is shown in Fig. 2.4. It will generally be necessary carefully to align the cell in the measuring beam, and to mask the beam to prevent radiation travelling through the walls (see Section 3.4). Micro cells are available with walls of black glass or black fused silica. These 'self-masking' cells do not require the use of a mask, but must be reproducibly placed in the beam if consecutive measurements are to be compared.

(f)   *Triangular cells*
A fluorescence cell that has three windows with one at 45° to the others, with the same overall dimensions as a Normal 10 mm cell, is becoming popular for the calibration of spectrofluorimeters. This is illustrated in Fig. 2.2.

(g)   *Tubular cells*
A number of clinical spectrophotometers and fluorimeters will accommodate test-tube-shaped cells. These are mentioned in this Volume because of their popularity, but are not recommended for accurate measurements.

(h)   *Demountable cells*
Cells for use with samples that contaminate the windows are more readily cleaned if the cell can be partially dismantled. This is particularly useful for short pathlength cells or flow cells with small chambers. One type of static flow cell that is commercially available is shown in Fig. 2.2, together with the spring-clip that holds the window in place. The whole assembly will fit a holder for Normal 10 mm cells.

2.1.3   *Grades of cells*
The Working Party has identified three grades of cell based on their dimensional tolerances:

(a)   *Grade A*
To be used for work of the highest accuracy. Such cells will be of the highest quality commercially available, and specially tested to ensure dimensional accuracy and optical quality. In some cases, e.g. 1 mm pathlength Normal cells, it is impracticable to achieve

sufficiently high standards in normal production, and the problems associated with the use of such cells means that they should not be used for measurements of the highest accuracy or reproducibility. Consequently, no Grade A specification is given for such cells.

(b)   *Grade B*

These are good quality cells for routine use. For precise absorbance measurements they should be calibrated against Grade A cells.

(c)   *Grade C*

These are the cheapest cells and are useful for teaching purposes, or as disposable cells for large-scale routine work and with difficult or dangerous samples.

2.1.4   *Special cells*

The cell types described in this section are considered to be those in most common use. Even amongst these common types, in some cases it is difficult to guarantee the dimensional tolerances and the cells are unduly expensive, and have therefore been omitted; for example, Grade A 1 mm cells were excluded for this reason. Such cells may be obtained from manufacturers by special order, but users are urged to consider alternative ways of making the measurement.

2.1.5   *High-accuracy cells*

Absorbance measurements of the highest accuracy demand knowledge of the cell pathlength. This can be achieved in one of two ways: (i) by asking the manufacturer to supply a cell that is very close to the nominal pathlength, or (ii) accurately measuring the pathlength of a standard production cell. To illustrate this, the selection of a Normal 10 mm cell required for high accuracy measurements will be considered. A Grade A cell (see Table 2.1) has a pathlength of 10.00 ± 0.01 mm, i.e. within 0.1 per cent. If required, a manufacturer could supply a cell with closer tolerances; this would probably be done by selection from a production batch of Grade A cells and would require a precision of measurement that would be difficult to achieve. The cell would thus be expensive, and so an alternative approach is recommended. The NBS measures the pathlength of production cells at ten points in the height of the cell. These can be purchased, together with the calibration certificate giving the pathlengths to the nearest $0.1\ \mu$m, i.e. 0.01 per cent of pathlength, and the relevant pathlength for the beam height in the user's instrument can be read off from the table. Descriptions of these cells and of the methods used are given by Mavrodineanu and Lazar[1, 2].

A similar conclusion is reached for short pathlength cells. The Grade B specification gives a pathlength of 1.00 ± 0.01 mm, i.e. within 1 per cent. In view of the difficulties associated with the use of cells of such short pathlength we do not recommend their use for measurements meriting a greater degree of accuracy than this. However, if more accurate measurements are to be carried out, we suggest that the best approach is to determine the pathlength of a production cell rather than purchase a cell especially made to closer tolerances. One manufacturer quotes the cost of a Normal 1 mm cell with a pathlength tolerance of $\pm 2\mu m$ as about ten times the cost of a production-run cell with Grade B tolerance.

2.2 Cell materials and assembly

2.2.1 *Window materials*
In choosing window materials, optical transmission in the relevant wavelength range is the prime consideration, although other factors must be taken into account, for example, the ability to seal the window effectively to the body of the cell; its rigidity, to avoid distortion; its hardness, to avoid scratching in use; its resistance to solvents and chemical attacks, and so on.

Five types of material are commonly used for cells in the UV and visible regions:

(a) *Synthetic fused silica*
In order to make silica free of the impurities found in natural quartz it is prepared by the high-temperature dissociation of purfied silicon compounds to give a material transmitting down to below 180 nm. It is virtually free of fluorescence.

(b) *Fused quartz*
Natural quartz can be fused to give a material that is effectively transparent down to 250 nm, but shows absorption below this because of metallic impurities. Fused quartz shows significant fluorescence.

(c) *Special optical glass*
This is an optical glass with high UV transmission. A cell made of such glass typically has a 75 per cent transmission point at about

310 nm. Borosilicate glasses have slightly better UV transmission characteristics than this, but are not often used for the manufacture of cells. Optical glasses soften and fuse at lower temperatures than fused silica, and are easier to work.

(d) *Glass.*
Windows of good quality optical glass transmit well throughout the visible region and down to about 350 nm in the near-UV region.

(e) *Plastics*
At the present time, disposable cells are moulded from polystyrene, polycarbonate and other transparent plastics which give low-cost cells of acceptable optical quality. The material is attacked by many organic solvents; the surface is readily scratched and can become coated by material from the sample solution. Newer materials have improved UV transmission.

Transmission curves for cells made from these materials are given in Section 2.2.3.

2.2.2 *Wall materials*

The walls of the cell need not be made from material of the same optical quality as the windows, except in the case of fluorescence cells. However, the thermal expansion coefficient should match that of the window material and, if the chemical resistance of the walls is different from that of the windows, the manufacturer must mark the cell accordingly.

2.2.3 *Window transmission*

The transmission of a cell can be measured in a spectrophotometer. The apparent fraction of the incident beam that reaches the detector is determined by:

(a) absorption by the window material;
(b) scatter and reflection at all four window surfaces;
(c) dispersion and deviation of the beam due to optical defects;
(d) fluorescence of the window material;
(e) cleanliness of the windows.

Cells made from synthetic fused silica, plastics and some glasses have low fluorescence, and so (d) is not significant. However, natural fused quartz can have strong fluorescence and a fused quartz cell when used in the UV region may emit visible light. Item (c) is not important when using good quality cells. With an empty cell, reflection at all four glass-air interfaces will limit the maximum transmission. Two of these reflections are reduced when the cell is filled, but (b) remains significant, for losses at the outer surfaces still limit the maximum transmission to about 92 per cent.

Figs 2.6 and 2.7 show typical transmission curves for cells made from different types of material. Procedures for measuring cell transmission are given in Chapter 9. The performance of cells in the UV region should be indicated by giving the wavelength at which the transmission of a clean cell filled with freshly-distilled water falls to 75 per cent, measured relative to the transmittance of air. This wavelength represents the lower limit to the spectral region in which the cell can be used satisfactorily; the cell can be used at shorter wavelengths, but with reduced accuracy.

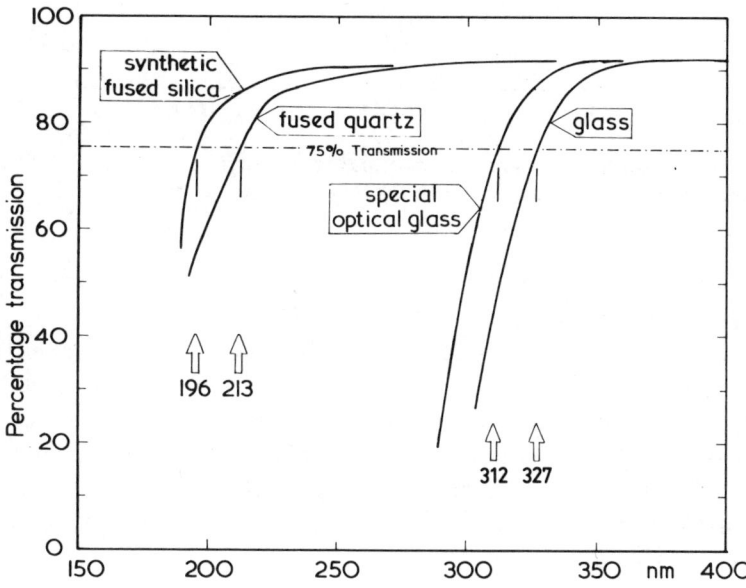

Fig. 2.6 *Typical UV transmission curves for cells with windows of synthetic fused silica, fused quartz, special optical glass, and glass. The cells were of 10 mm pathlength and filled with freshly-distilled water. The arrows indicate points of 75 per cent transmission. The poor transmission of the water below 200 nm means that the 75 per cent point of the synthetic fused silica cell is at a longer wavelength than for an equivalent dry cell.*

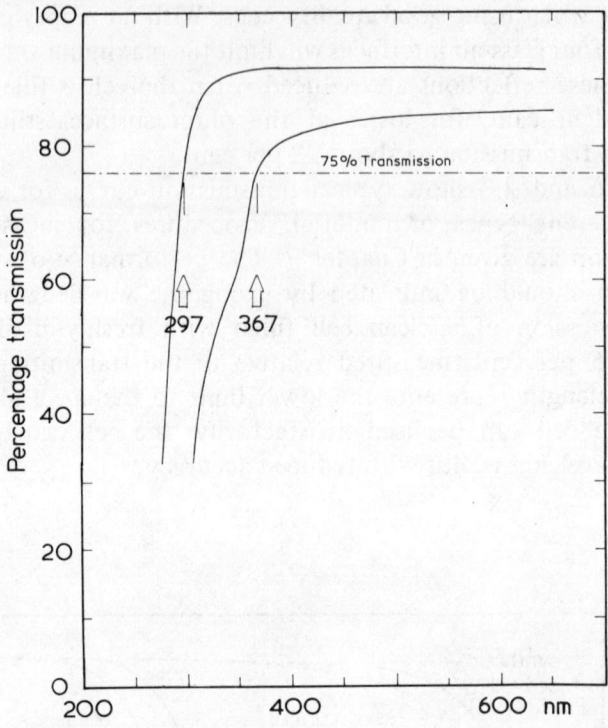

Fig. 2.7   *Transmission curves for two makes of plastic cell filled with freshly-distilled water. The cells appear to be made from different plastics.*

2.2.4   *Methods of construction*

(a)   *Fused construction*

Most cells that are sold today have walls made from the same type of material as the windows, and all parts of the cells are joined by fusing. While this may cause slight distortion at the edges of the windows, modern techniques reduce this to a minimum, and the resulting cell is strong, as resistant to thermal shock as the material of construction, resistant to attack by cleaning agents and solvents, and is easy to clean.

(b)   *Sintered construction*

Some cells are made by using a glass material which has a melting point lower than that of the walls and windows interposed between the components. The assembly is then heated to sinter the joints together. The lower temperatures used means that there is no distortion of the windows, but the joints may not be as strong as fused ones, are liable to be pin-holed, and are generally less resistant to chemical attack. Manufacturers should mark cells that are of sintered construction.

(c) *Cemented construction*

In some cases, for example in complex flow cells where fusion of all joints is impossible, some components are joined with a cement which gives a joint that may be less strong mechanically, less resistant to chemical attack, and more liable to contamination by absorbed materials than the window material. Manufacturers should mark cells of cemented construction.

(d) *Demountable cells*

Cells made with removable windows may be readily cleaned and, if necessary, the windows replaced. Carefully finished surfaces to the cell body, an efficient clamping system and absolute cleanliness are essential to guarantee reproducible assembly of the cell and freedom from leaks. An advantage of this construction is that the body and windows can be of widely differing materials; for example, special flow cells for chromatograph monitoring are made with a stainless steel body and fused silica windows sealed with PTFE washers. Since the PTFE is relatively rigid, the cell can be reassembled with a reproducible pathlength, though a strong clamping device is necessary to prevent leakage (see Fig. 2.3).

2.2.5 *Fluorescence of material*

The fluorescence of windows can cause errors in both absorption and fluorescence spectroscopy. For example, if a sample with high absorbance in the UV is being measured at short wavelengths, fluorescence of the entrance window of the cell, which will be emitted at a wavelength longer than that of the measuring beam, may not be absorbed by the solution and so may constitute a significant fraction of the radiation reaching the detector. This problem is more acute in fluorescence measurements.

The structure of fused quartz has network defects, generally associated with reduced metallic ions, that give rise to fluorescence at about 565 nm. Synthetic fused silica is virtually free of metallic impurities and shows negligible fluorescence.

Some glasses show pronounced fluorescence and this, coupled with their UV absorption, makes them unsuitable for most fluorescence measurements.

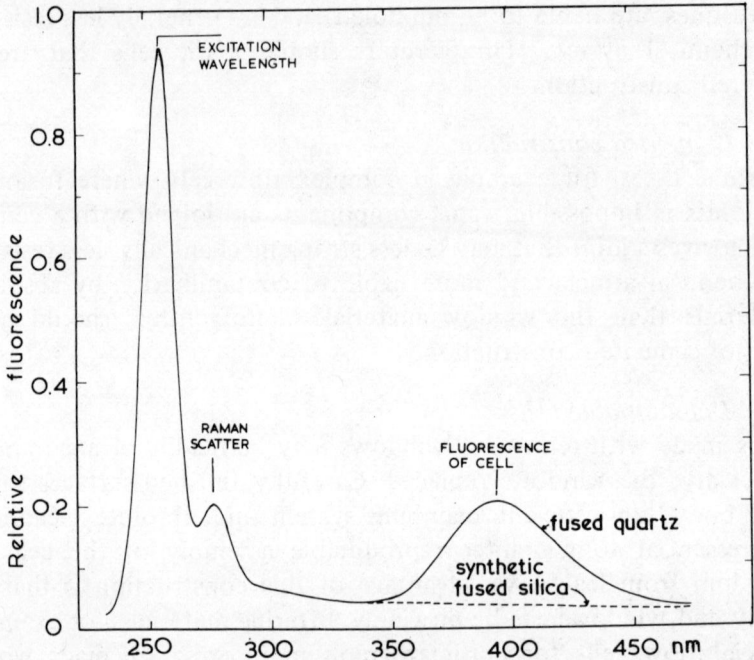

Fig. 2.8   *Typical fluorescence emission spectra for cells of synthetic fused silica and fused quartz. The cells were 10 mm square internally, and filled with freshly-distilled water. The spectra were recorded with a Perkin-Elmer Model 204A fluorescence spectrophotometer using 10 mm entrance and exit slits and an excitation wavelength of 250 nm.*

Typical fluorescence spectra of cells filled with water and measured in a right-angle spectrofluorimeter are given in Fig. 2.8.

2.2.6   *Optical specification of the cell*

Several criteria beyond window transmittance have to be considered in assessing the optical quality of a cell:

  (a) closeness to specified pathlength;
  (b) parallelism of the inner faces of the windows;
  (c) parallelism of the faces of each window;
  (d) flatness of the windows over the working area;
  (e) freedom from surface blemishes over the working area;
  (f) quality of surface polish over the working area;
  (g) parallelism of the outer faces of the windows.

Items (a) and (g) can be measured accurately by mechanical means for open types of cell, though short pathlength cells and cylindrical cells present special problems. Methods for the measurement of pathlength are described in Section 9.1. Items (b) and (c) can also be measured by mechanical methods or, since these defects cause a displacement or deviation of the beam, an optical test can be used, such as those given in Section 9.4.

Item (d) is difficult to check, particularly for the inner faces of windows. However, such defects will cause scattering or dispersion of the beam, and a test for this is given in Section 9.4.

Items (e) and (f) can be assessed by visual inspection but will also have an effect upon the measured transmission. Thus a cell that has a poorer transmission than expected, for the material of its construction, may well have surface defects. It is important to ensure that cells are properly cleaned before any transmission measurement is made, for contamination can have a major effect upon the UV transmission.

2.3 Cell design

2.3.1 *Construction methods*

(a) *Fused construction*
The most satisfactory cells are formed from plates or plates and tubing fused together. The windows are first polished and then fused to the body. Modern production methods minimize the distortion of the windows and body during the process, but some loss of optical quality around the edges of the windows and slight variability in the pathlength of the cell is inevitable. In addition, fusing quartz and synthetic silica can cause a deposit of amorphous silica to form on the surrounding cold surfaces.

In Normal cells and in good quality Semi-micro and Micro cells, distortion of the windows is unlikely to affect the working area as it is specified in Table 2.1. However, Micro cells are sometimes used with beams that fall outside the working area, and then such distortions might have an effect. In Semi-micro and Micro cells made with thick walls, it is imperative that the fusing is carried out effectively, and that there are no gaps inside the cell where the windows meet the side walls.

(b) *Cemented and sintered construction*
If the windows and walls of the cell are not heated to fusion point during assembly, there will be little distortion. However, the cells are less satisfactory for the reasons given in Section 2.2.4.

Table 2.1: *Rectangular absorption cells*

Cell type	Pathlength	Working area		Pathlength (d) ± tolerance			External length (e)		Max. internal width (b)	Angular deviation of beam		
		Min. height (f) (i)	Min. distance sides to C.L. cell (i₁ and i₂)	A	B	C	Max.	Min.		A	B	C*
Normal	1	22	3.5	—	0.01	—	3.55	3.10	10.0	—	5'	—
	2	22	3.5	—	0.01	—	4.55	4.10	10.0	3'	5'	—
	5	22	3.5	0.01	0.02	—	7.55	7.10	10.0	3'	5'	—
	10	22	3.5	0.01	0.04	0.30	12.55	12.10	10.0	3'	5'	10'
	20	22	3.5	0.02	0.10	—	22.55	22.10	10.0	3'	5'	—
	40	22	3.5	0.03	0.10	—	42.55	42.10	10.0	3'	5'	—
Semi-micro	5	10	1.75	0.01	0.02	—	7.55	7.10	4.3	4'	6'	—
	10	10	1.75	0.01	0.04	0.30	12.55	12.10	4.3	4'	6'	10'
	20	10	1.75	0.02	0.10	—	22.55	22.10	4.3	4'	6'	—
Micro	5	8	1.0	0.01	—	—	7.55	7.10	2.3	5'	—	—
	10	8	1.0	0.01	—	—	12.55	12.10	2.3	5'	—	—
	20	8	1.0	0.02	—	—	22.55	22.10	2.3	5'	—	—

All dimensions in millimetres

The recommended height of the centre of the working area above the base (h) should be 15 mm in all cases.

The maximum height of the cell body (c) is 45 mm, and for stoppered cells, the height including the stopper should not exceed 55 mm.

The minimum distance between the top of the working area and the top of the chamber should not be less than 5 mm.

The external width is 12.45 ± 0.15 mm for Grade A and B cells, and 12.40 ± 0.30 mm for Grade C cells.

*These figures are based on the tolerances of current plastic cells and may require revision as manufacturing techniques improve.

Table 2.2: *Cylindrical absorption cells*

Preferred pathlengths	Working Area	Pathlength (d) ± tolerance	External length (e)		Max. external diameter (n)	Angular deviation of beam
	Min. diameter (p)		Max.	Min.		
50	16	0.02	52.55	51.0	22.5	5'
100	16	0.02	102.55	102.0	22.5	5'

All dimensions in millimetres

The maximum overall height is 40 mm, including stoppers.

(c) *Moulded cells*

Plastic cells are generally moulded in one piece. The inner part of the mould must be tapered and so the windows are not parallel. The tolerances of the cell and the optical quality of the window are determined by the quality of the mould, and there are considerable differences in the cells from different manufacturers.

It is hoped that moulding techniques will be developed for glass and silica materials in order to produce cheaper cells for routine and disposable applications.

2.3.2. *Dimensional tolerances*

The preferred pathlengths, dimensions and tolerances for rectangular and cylindrical absorption cells are given in Tables 2.1 and 2.2. The pathlength of the largest rectangular cell is taken as 40 mm rather than 50 mm as recommended in the DIN specification for three reasons:

(a) beam divergence in many instruments means that the beam size would exceed the working area over a 50 mm path;

(b) since they will probably be used only with samples too dilute to measure in a 20 mm cell, a factor of 2 is more convenient in scaling down the result;

(c) the cell-holders of some instruments will not accept 50 mm cells.

Only two pathlengths are proposed for cylindrical cells as it is only necessary to use these when long pathlengths demand a large working area. A specification for tubular cells is given in Table 2.3.

Three types of fluorescence cell are given in Table 2.4. Other sizes are in use, e.g. 7 x 7 mm, but the adoption of the three sizes listed in Table 2.4 is recommended.

Table 2.3: *Tubular absorption cells*

Internal diameter	Internal diameter ± tolerance	External diameter	
		Max.	Min.
10	0.5	12.55	12.10

All dimensions in millimetres.

The base may be hemispherical or flattened, but the dimensions of the cell should fall within the above limits from a point 7 mm above the lowest point of the outside surface of the cell.

It must again be stressed that the tolerances given in the tables were chosen to be adequate for most purposes at a reasonable cost. Cells can be made to tolerances better than these, and most manufacturers will be able to supply such as special items.

2.3.3 *Ease of handling, cleaning and emptying*

The dimensional specifications given above allow some variation in the shapes of cells, but other factors should also be taken into consideration:

(a) A slight rounding or chamfer of the external edges is desirable, but since in many cell holders the cell is located by contact near the junction of windows and walls, excessive rounding will cause poor alignment. Fig. 2.9 shows the recommended style of chamfer;

(b) A flat base for rectangular cells will allow them to stand safely on the bench;

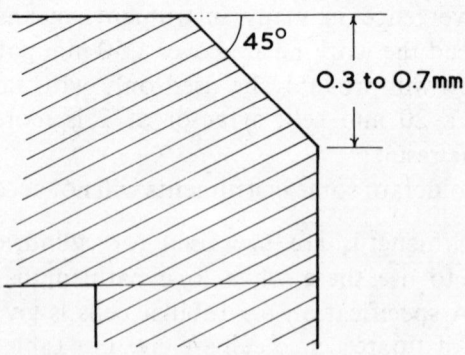

Fig. 2.9 *Diagram illustrating the recommended chamfer for the external edges of cells.*

Table 2.4: *Rectangular fluorescence cells*

Cell type	Pathlength	Working area		Pathlength 1 (b) ± tolerance		Pathlength 2 (d) ± tolerance		External dimensions (a) and (e)	
		Min. height (f)	Min. distance to sides to C.L. cell (j_1 and j_2)	B	C	B	C	Max.	Min.
Normal	10	22	3.5	0.05	0.30	0.05	0.30	12.55	12.10
Semi-micro	5	10	1.75	0.02	0.20	0.02	0.20	7.55	7.10
Micro	3	8	1.00	0.02	0.20	0.02	0.20	5.55	5.10

All dimensions in millimetres

The recommended height of the centre of the working area above the base (h) should be 15 mm in all cases.
The maximum height of the cell body (c) is 45 mm, and for stoppered cells the height including the stopper (g) should not exceed 55 mm.
The minimum distance between the top of the working area and the top of the chamber should not be less than 5 mm.

(c) The internal corners should be smooth and rounded if possible; there should be no crevices to retain solutions or cleaning agents;

(d) The stoppers should be as large as possible to facilitate filling and emptying. While it is recommended that the socket for the stopper is in the form of a block (see Section 2.3.4) this should be sealed on so as to minimize the amount of liquid trapped in the corners at the top of the cell when it is inverted;

(e) Tall rectangular cells have the advantage of being easy to remove from the cell holder, but shorter cells are more economical in materials, are easier to clean, and have a lower centre of gravity.

2.3.4 *Socket and stopper design*

Sockets and stoppers of the 'bottle top' design, specified in BS 3875, should be fitted to short pathlength cells (i.e. less than 5 mm) as they give an opening that is relatively large compared to the chamber of the cell, for ease of filling and emptying. However they are rather fragile and so it is recommended that longer pathlength rectangular cells are fitted with 'block-top' stoppers, as illustrated in Fig. 2.1. The stoppers should have a standard taper of 1 in 10, should be of a standard size, and should be made of the same material as the cell body or of PTFE. In general, PTFE is preferable as it will not seize. A sufficient length of stopper should project above the block for its easy removal, and should be shaped so that it can be grasped with the finger tips, or with pliers if necessary. It should be borne in mind that PTFE has a much higher coefficient of thermal expansion than glass or fused silica. Since the main function of the stopper is to prevent loss of liquid by spillage or evaporation, it should be air-tight.

2.3.5 *Fluorescence cell design*

The design considerations for absorption cells apply equally to cells used for fluorescence measurements. However, one advantage of the fluorescence technique is that the accuracy of the results is less dependent upon the optical quality of the cells than in absorption photometry. This is because most commercial spectrofluorimeters are arranged so that only a small central section of the cell is viewed by the detector. Consequently tubular cells may be used, though it is important that these have a uniform wall thickness, as this affects the 'lens action' of the cell.

The material of construction for a fluorescence cell must transmit both the excitation and emission wavelengths and, since the former is often below 300 nm, glass and plastics are not suitable. Every cell should be checked for intrinsic fluorescence, as variations are found between batches of the same material, particularly in the case of natural quartz. Cemented construction should be avoided since many cements fluoresce.

2.3.6 *Sampling cell design*

There are many patterns of sampling cell in use, but the pattern shown in Fig. 2.1 is recommended, which has the specification of a Semi-micro cell and will fit in a Normal cell-holder. A 10 mm path-length cell requires a volume of 1.2 ml to fill it to a level of 5 mm above the working area. The size of the filling tubes is not critical for the sampling mode of operation.

2.3.7 *Flow cell design*

A typical flow cell is shown in Fig. 2.2. In this case the dimensions are those of a rectangular Micro cell although, for the greatest efficiency of scavenging and the best time resolution in on-line operation, the volume must be further reduced, with a resulting reduction in working area. The measuring beam must then be suitably masked.

The tubing and connection to the cell should be of the minimum practicable volume; proposals for the connection of tubing are illustrated in Fig. 2.10.

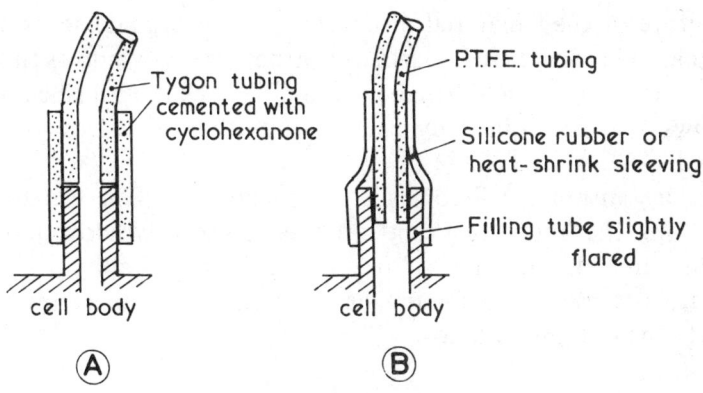

Fig. 2.10 *Practical methods for the connection of plastic tubing to sampling and flow cells. A: A seal is made with a piece of larger tubing cemented to the end of the supply tube; B: The cell-filling tube is specially flared to allow the PTFE tubing to make a push-fit. The outer sleeve gives mechanical stability to the connection.*

In high-pressure liquid chromatography, the highest resolution is required and cells with a 1 mm diameter working area are used - a 10 mm pathlength cell then has a volume of about 8 μl. Special consideration must be given to the optical design of the measuring instrument used with such a cell.

2.3.8   *Labelling of cells*

Three pieces of information should be indelibly marked on the cell in a prominent position outside the working area:

(a) The 75 per cent transmission point in the UV, measured as detailed in Chapter 9. This should be given as a 3-digit number, to the nearest 5 nm. The DIN specification recommends that both UV and IR limits shojld be given to the nearest 10 nm, but we feel that the IR performance up to 1000 nm is not critical for most measurements and is rarely an important factor in choosing a cell. On the other hand, a precise 75 per cent point in the UV is important in the selection of a cell and also gives a good indication of the nature of the window material.

(b) Whether materials differing in nature from the window material have been used in the construction, for example, glasses of different chemical or physical properties for the walls, or cements or different glasses for the joints. Plastic stoppers or lids are not considered to be part of the cell in this context.

Cells of homogeneous construction, i.e. fused or moulded, should be labelled 'S', 'G' or 'P' (for fused silica, glass or plastic) to indicate the nature of their material of construction as a guide to usage and cleaning. Cells with fused quartz or synthetic silica windows that have glass components should be labelled 'S-G' and cells with silica or glass windows that are joined by cements or other adhesives should be labelled 'S-CEM' or 'G-CEM'.

(c) The entrance window should be marked. This can be done by putting the above markings on the entrance window above the working area. In the case of plastic cells this is impracticable and cells are often marked on the base. In this case, an arrow should indicate the entrance window:

In the case of flow and sampling cells, the entry tube of the cell should be marked by an arrow pointing in the direction of flow, placed on the cell body as near ass possible to the point at which the tube joins the body. If this is impractical, the tube itself should be marked with an indelible band.

Other information, for example the maker's name, can be marked on any face of the cell so long as it does not interfere with the working areas of the windows

References

1 Mavrodineanu, R. and Lazar, J.W. (1973), *Clin. Chem.*, **19**, 1053.
2 *Standard Reference Material 932 - Quartz cuvette for spectrophotometry*, NBS Special Publ. 260-32 (obtainable from Superintendent of Documents, U.S. Govt Printing Office, Washington DC 20402).

3 Instrument design considerations

3.1 Introduction

This chapter considers features of the design of spectrometers that have a direct bearing upon the choice and use of cells. Information has been obtained by examining current production instruments from six spectrophotometer manufacturers: Bausch & Lomb, Beckman, Hitachi, Perkin-Elmer, Pye Unicam and Varian Techtron. All of the instruments examined accommodate 10 mm Normal cells conforming to BS 3875.

3.2 Beam dimensions

The light transmitted through a spectrophotometer cell is usually obtained from a grating or prism monochromator. The focus of the light is therefore usually the image of the exit slit of the monochromator. In order that the light passes through the cell with maximum clearance, the cell is usually placed near this focus. The shape of the light beam going through a cell is, therefore, approximately an upright narrow rectangle.

Half of the above instruments have beams centred 15 mm above the bottom of the cell, while half have beams centred at 10 mm. In the interests of standardization, it is recommended that instruments should have beams centred 15 mm above the bottom of the cell. Cells designed for a beam 10 mm above the bottom may easily be used in an instrument with a 15 mm beam position by the insertion of a spacer beneath the cell. A 15 mm beam position also allows the accommodation of such things as magnetic stirrers in the cell below the beam, together with a miniature stirrer driver underneath the cell.

3.3 Beam divergence

Among the instruments examined, the maximum angle of any ray in the light beam from the axial ray was 5°. For a beam of uniform radiance at all cross-sections, this would produce an error in the absorbance reading of +0.2 per cent. This compares with ±0.5 per cent error due to pathlength in a Grade B cell, and ±0.5 per cent absorbance accuracy, typical of many spectrophotometers. It is recommended that this maximum figure of 5° should not be exceeded, which also conforms to the requirements of the International Committee on Illumination [1].

A beam of 5° divergence with an image 0.25 mm wide at the focus will pass through the 2 mm-wide working area of a 20 mm Micro cell without obstruction; it will, therefore, pass readily through most other types of cell.

The beam divergence frequently prohibits the use of cells with pathlengths longer than 40 mm since the beam cross-section becomes more square away from the focal plane of the slit image. It is recommended therefore that rectangular cells do not exceed 40 mm, and cylindrical cells - better suited to a square beam cross-section - be used for pathlengths of 50 and 100 mm. Beam divergence similarly prohibits the use of long Micro cells, so we have recommended a maximum of 20 mm pathlength for them.

3.4 Beam masking

When using micro cells in some instruments, it will be found that part of the beam falls outside the working area of the cell, or that divergence or misalignment of the beam after it enters the cell causes it to be reflected off the inner faces of the walls. To overcome these effects, the user must arrange a mask to limit the size of the beam.

It is essential, particularly when working with highly absorbing samples, that the measuring beam cannot by-pass the sample by travelling through the walls of the cell; for example, if 1 per cent of the measuring beam by-passes a sample of absorbance $2A$, the apparent absorbance will be 15 per cent below the true value. The choice of a suitable mask is a compromise arrived at by consideration of the cell dimensions and the optical properties of the instrument. The aperture in the mask should be the same size as, or smaller than, the working area of the cell. It is normally placed in front of the entrance window. Ideally, it should be rigidly attached to the instrument, so that movements of cell-holder or cell will not affect the instrumental baseline. However, in practice, it is often attached to the cell-holder, and care should then be taken that the cell-holder can be reproducibly replaced in the instrument. Another alternative is the use of 'self-masking' Semi-micro cells in which the walls are made of black glass or fused quartz. In this case, care must be taken that the cell is precisely located in the cell-holder and that the beam cannot be reflected off the inner faces of the walls. It is also essential that the cell is located so that the exit window is near the focus of the beam, i.e. the beam is converging as it passes through the cell.

The alignment of the mask, or the self-masking cell, should be checked by visual inspection with the wavelength set at 550 nm and the slits opened wide. Most of the beam should pass through the aperture and, if not, adjustments of the mask, cell or cell-holder must be made. The effect of closing the slit should be noted - the reduction in the size of the beam, if any, will depend upon the design of the instrument.

The performance of the instrument should then be checked over the operating wavelength range - if the aperture is too small, the instrument may become excessively noisy, or it may become impossible to reach the absorbance zero. In double-beam instruments it may be necessary to attenuate the reference beam in order to get a satisfactory beam balance.

In general, such problems will only be encountered when using Micro cells; the dimensions of Semi-micro cells given in Chapter 2 will allow their use under normal operating conditions without special masking.

3.5   Cell-holders

All the cell-holders examined could accommodate a Normal 10 mm cell. Some instruments have special holders for cylindrical cells and special cells. There is usually a large tolerance on the length, but the cell width may be limited by the width of the holder. A maximum cell width of 12.6 mm has, therefore, been recommended for all rectangular cells.

For accurate re-location of a rectangular cell in a cell-holder, which is necessary for Semi-micro and Micro cells, a corner locating mechanism is required. While present instruments vary in the means of locating the cell in the corner (a single spring or two perpendicular springs), it is recommended that manufacturers should be encouraged to take the same side of the cell as the reference plane, namely the side to the right of the beam looking in the direction of the beam, towards the detector. This means that the external width of the cell is less important, and the thickness of only one wall is critical, hence a smaller internal width for a given working area may be achieved. Due to the problem of friction, care should always be taken when inserting a cell to ensure that it is properly located. Good contact with the cell-holder also ensures better temperature control of the cell. The cell-holder should be suitably recessed so that the working area cannot be scratched as the cell is inserted or removed.

Errors of measurement can arise through mechanical defects of the cell-holders:

(a) If the cell is not located squarely in the cell-holder or the cell-holder is not square to the measuring beam, the beam will be displaced on passing through the cell, and the apparent pathlength of the cell will be increased. BS 3875 recommends that the orientation of the face of the cell should be controlled to within 3′ in any direction, though in practice this would be difficult to measure. An alignment error of 30′ in a 40 mm cell will displace the beam by 0.1 mm, which in most instruments will not cause any serious error.

(b) It should be possible to replace the cell reproducibly in the cell-holder. Ingle [2] has tested two simple commercial spectrophotometers and shown that, for both, the most serious loss of precision was due to poor cell positioning. Difficulties may arise through design faults in the cell-holder, weakening or corrosion of the locating springs or a build-up of dirt in the cell-holder. A cell with chipped corners or which is under-sized may not locate properly. The reproducibility of the alignment of cells can be checked by attaching to or inserting into a test cell a mask with a vertical slit 1 mm wide. The monochromator is turned to 550 nm, the slits of the instrument opened, and visual inspection made to ensure that most of the light passes through the aperture in the cell. The apparent transmission is measured, the cell removed, replaced and measured again. This is repeated several times. The measured transmissions should be within 1 per cent of their mean value.

If this cannot be achieved, check the reproducibility of the instrument itself by repeating the measuring routine several times without the cell in the holder.

For measurements of the highest precision, it is recommended that a special cell-holder is made so that the cell is clamped firmly in

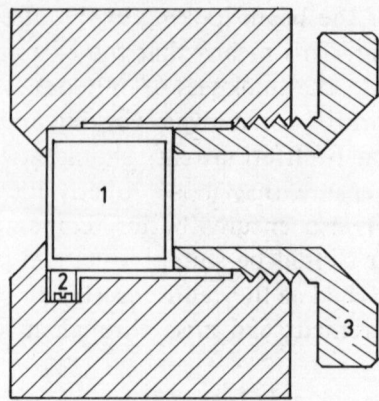

Fig. 3.1   *Horizontal section of a special cell-holder with clamping screw used in the Radiation Dosimetry Laboratory, NPL. 1: 10 mm rectangular cell; 2: Leaf-spring ensuring lateral location of the cell; 3: Clamping screw.*

position by a screw device. The cell is then clamped in position for the duration of the series of measurements. Fig. 3.1 shows a satisfactory design for a cell-holder which locates a Normal 10 mm cell by means of a screw clamp.

(c) The cell-holder should re-position accurately when it is removed from the instrument or, if it is on a slide or a turret, when it is moved from one position to nother and back again. This can also be checked using the aperture cell described above.

(d) The positions of a multi-cell-holder should be equivalent. Again, the apparent transmission of the aperture cell should be the same when it is placed in the different positions.

3.6   Detectors

The response of photocells and photomultipliers is sensitive to the position at which the beam enters the window. This positional sensitivity depends on the detector design, the quality of the photo-sensitive coating, the wavelength and the surface of the detector. Vertically mounted side-window photomultipliers are particularly sensitive to horizontal variation in the position of the beam. Care must be taken to ensure that the beam moves as little as possible and in a reproducible manner when the cell is introduced. This means that the cell must have its windows perpendicular to the beam, and the surfaces of the windows must be as parallel as possible. It must be remembered that plastic cells will cause a significant vertical deviation since they are tapered as a result of the moulding process.

The quality of cells and detectors are to some extent complementary: a poor cell can give satisfactory results if the detector has a uniform sensitivity over its surface. If the detector in an instrument is changed, it is advisable to check the performance of cells and cell-holders used for the most accurate measurements.

References

1   Commission Internationale de l'Eclairage, Publication 15.
2   Ingle, J.D., (1977), *Anal. Chim. Acta,* **88**, 131.

4 Liquid absorbance standards

4.1 Introduction

The advantage of using a solution as an absorbance standard is that the procedure for its measurement closely resembles that for a normal sample. The use of solution standards combines both operator error and cell errors in one set of measurements and, providing that Beer's Law is obeyed for the standard concerned, can be used at any absorbance level in the instrument's range. This method utilizes a reference blank and is also ideal for calibrating flow systems as it reproduces cell dimensions precisely [1]. The disadvantages of solution standards are that they do not possess the high degree of optical neutrality found with solid filters, they require careful preparation and they usually have larger temperature coefficients than solid standards.

4.2 Standards for the 200 - 400 nm region

4.2.1 *Potassium dichromate*

Of all the compounds suggested for the preparation of absorbance standards, potassium dichromate has been the most extensively studied. The choice of solvent conditions is the point most in dispute, for both acidic and alkaline aqueous media have their advantages and disadvantages.

(a) *Potassium dichromate in acid solution*
The NBS has reported detailed studies on potassium dichromate in

acidic media [1-4]. In solution the following equilibrium processes occur:

$$H_2CrO_4 \xrightleftharpoons{K_1} H^+ + HCrO_4^- \qquad K_1 = 0.16 \text{ [5]}$$

$$HCrO_4^- \xrightleftharpoons{K_2} H^+ + CrO_4^{2-} \qquad \begin{aligned} K_2 &= 3.2 \times 10^{-7} \text{ [6]} \\ &= 3.0 \times 10^{-7} \text{ [7]} \end{aligned}$$

$$2HCrO_4^- \xrightleftharpoons{K_3} Cr_2O_7^{2-} + H_2O \qquad \begin{aligned} K_3 &= 43.7 \text{ [5]}; 35.5 \text{ [8]}; \\ &\quad 33.0 \text{ [9]}; 32.9 \text{ [2]}. \end{aligned}$$

The dissociation constants given above were all measured at 25°C. At pH = 10, the chromium (VI) present exists as 99.9 per cent CrO_4^{2-}, whereas in weakly acidic solution (approximately pH =3) the predominant species is $HCrO_4^-$ with less than 0.1 per cent contribution from H_2CrO_4. As indicated above, $HCrO_4^-$ is capable of dimerization and it is this process that causes difficulty when measuring the absorption spectrum under acid conditions. The amount of dimer formed is dependent on the initial concentration of the salt present: at 1.36×10^{-4} M, 0.9 per cent dimer is formed; while at 6.80×10^{-4} M, 4.2 per cent dimer is present [3].

Fig. 4.1 shows the absorption spectrum at a concentration of 1.0×10^{-4} M, while Fig. 4.2 shows that the molar absorptivity of the dimer is greater than that of $HCrO_4^-$ at most wavelengths, and so it is the presence of dimer that causes the observed deviations from Beer's Law [2, 10]. This is the main drawback to the use of acid solutions as a standard.

The choice between the use of sulphuric acid or perchloric acid as solvent is a less important issue. The NBS points out that 0.01 N H_2SO_4 has two disadvantages compared with perchloric acid at pH = 3: sulphuric acid has a greater ionic strength and hence greater salt effects, and there is the possibility of forming mixed chromium (VI)-sulphato complexes [3].

Since the values of the dissociation constants K_n are dependent upon the acidity of the solution [8], when using 0.01 N H_2SO_4 at pH = 2 there is an increased possibility of deviations from Beer's Law. At this pH it is possible for H_2CrO_4 to be present as well, and this causes even more uncertainty in the observed absorbance values. However several workers prefer sulphuric acid, as it is safer to handle and more readily available. In general it is accepted that a pH of 3 is preferable to a pH of 2, but it should not be greater than 3. Whichever is used, it is essential that the nature and concentration of the

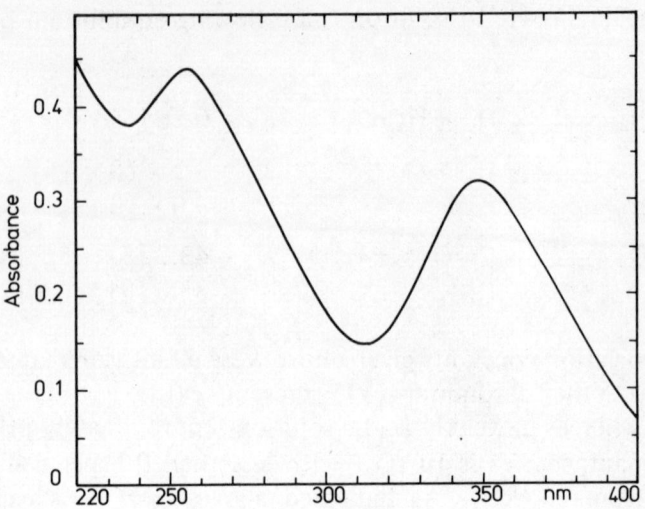

Fig. 4.1   *Absorption spectrum of 1.0 × 10⁻⁴ M potassium dichromate in 0.001N perchloric acid measured in a 10 mm cell. Temperature and pH unspecified. Redrawn from [3].*

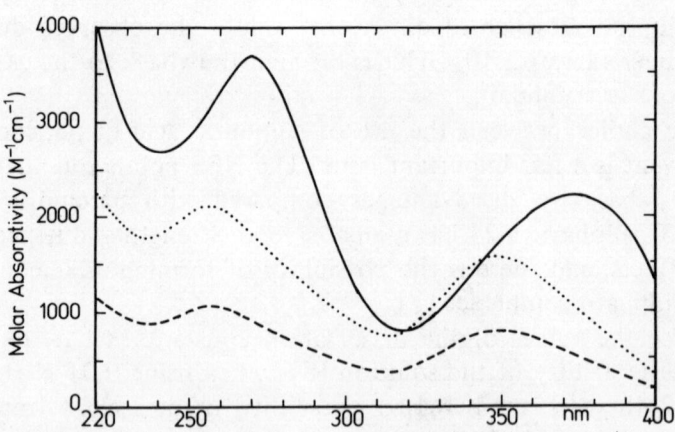

Fig. 4.2   *Computed absorption spectra for the chromate (- - - -) and dichromate (———) ions in perchloric acid, pH = 3.0. The curves were derived by extrapolation from the spectra of dichromate solutions of various concentrations. The spectrum of the chromate ion multiplied by 2 is also shown (. . . .). At the isobestic points of this and the dichromate curve the solution should obey Beer's Law. Redrawn from [3].*

solvent is quoted, that the pH of the solution is accurately measured and quoted, and the concentration of the potassium dichromate also given.

The molar absorptivity of potassium dichromate in acid solution is sensitive to temperature. At 313 nm the temperature coefficient is 0.02 per cent per °C, while at 235, 257 and 350 nm it is − 0.05 per cent per °C over the range 20-30°C [2]. The photochemical stability of the solution is high, there being no significant changes in absorbance on prolonged exposure to a 250 watt high-pressure mercury arc lamp [11].

Results obtained in various measurements and collaborative trials are summarized in Table 4.1. A survey by Beeler [12] has been

Table 4.1: *Molar absorptivity values for potassium dichromate in acidic solutions. All solutions in 0.01 N H_2SO_4 unless specified otherwise.*

Concentration	Instrument	Number of instruments	Molar absorptivity (M^{-1} cm^{-1})				Ref.
			235 nm	257 nm	313 nm	350 nm	
0.04902 g l^{-1}	Beckman	38	3670	4277	1438	3141	40
0.04902 g l^{-1}	Unicam	20	3224	4236	1427	3132	40
0.04902 g l^{-1}	Uvispek	14	3669	4261	1429	3120	40
0.04902 g l^{-1}	NPL special	1	3679	4279	1428	3152	40
0.05 g l^{-1}	Beckman	48	3646	4238	1432	3152	41
0.05 g l^{-1}	Unicam	29	3657	4244	1439	3150	41
0.05 g l^{-1}	Beckman (W Lamp)	48	–	–	–	3196	41
0.05 g l^{-1}	Zeiss Opton	1	3721	4226	1481	3189	41
0.05 g l^{-1}	Cary	1	3496	4107	1394	3072	41
0.05 g l^{-1}	Beckman & Cary	NS	–	4231	–	3145	42
0.05 g l^{-1}	Bausch & Lomb	NS	–	4231	–	3154	42
0.05 g l^{-1}	NBS special	NS	–	4230	–	3150	3
0.06 g l^{-1}	NS	NS	3666	4251	1427	3160	43
0.06 g l^{-1}	NS	NS	3657	4275	1436	3157	43
0.06 g l^{-1}	NS	NS	3660	4272	1436	3174	43
0.06 g l^{-1}	NS	NS	–	4242	1424	3145	43
0.06 g l^{-1}	NS	NS	3642	4245	1421	3133	43
0.06 g l^{-1}	NS	NS	–	4236	1412	3130	43
0.06 g l^{-1}	NS	NS	3651	4251	1430	3142	43
0.06006 g l^{-1}	Cary 14	9	3695	4245	1442	3148	44[a]
0.06006 g l^{-1}	Cary 16	11	3695	4257	1442	3148	44[a]
0.09117 g l^{-1}	Beckman	38	3700	4292	1437	3171	40
0.09117 g l^{-1}	Unicam	20	3685	4267	1434	3142	40
0.09117 g l^{-1}	Uvispek	14	3637	4301	1438	3134	40
0.09117 g l^{-1}	NPL special	1	3727	4340	1436	3178	40
0.1 g l^{-1}	Beckman	48	3672	4264	1433	3163	41
0.1 g l^{-1}	Unicam	29	3693	4305	1436	3162	41
0.1 g l^{-1}	Beckman (W Lamp)	48	–	–	–	3153	41
0.1 g l^{-1}	Zeiss Opton	1	3722	4257	1470	3163	41
0.1 g l^{-1}	Cary	NS	3633	4242	1412	3127	41
0.1 g l^{-1}	Beckman & Cary	NS	–	4254	–	3154	42
0.1 g l^{-1}	Bausch & Lomb	NS	–	4251	–	3160	42
0.12002 g l^{-1}	Cary 14	9	3719	4307	1433	3166	44[a]
0.12002 g l^{-1}	Cary 16	11	3716	4307	1439	3171	44[a]

Table 4.1 Cont'd.

Table 4.1 Cont'd.

Concentration	Instrument	Number of instruments	Molar absorptivity (M^{-1} cm^{-1})				Ref.
			235 nm	257 nm	313 nm	350 nm	
NS	NS	NS	3677	4266	1441	3148±29	20[b]
0.05 g kg^{-1}	Cary	5	–	4196	–	3113	33
0.05 g kg^{-1}	Bausch & Lomb	1	–	4161	–	3072	33
0.05 g kg^{-1}	Bausch & Lomb	1	–	4238	–	3108	33
0.05 g kg^{-1}	Zeiss PMQ-11	1	–	4173	–	3096	33
0.05 g kg^{-1}	Beckman DU	20	–	4167	–	3096	33
0.05 g kg^{-1}	Beckman DB	14	–	4144	–	3049	33
0.05 g kg^{-1}	Beckman DK	5	–	4179	–	3108	33
0.05 g kg^{-1}	Cary	2	–	4185	–	3108	33
0.05 g kg^{-1}	Perkin-Elmer 101	1	–	4114	–	3078	33
0.05 g kg^{-1}	Perkin-Elmer 111	1	–	4114	–	3166	33
0.05 g kg^{-1}	Coleman-Hitachi 124	2	–	4314	–	3125	33
0.05 g kg^{-1}	Hitachi-Perkin-Elmer 139	3	–	4073	–	3090	33
0.05 g kg^{-1}	Perkin-Elmer 202	1	–	4002	–	2943	33
0.05 g kg^{-1}	Perkin-Elmer 350	1	–	4485	–	3178	33
0.05 g kg^{-1}	Perkin-Elmer 402	1	–	4061	–	2884	33
0.05 g kg^{-1}	Bausch & Lomb	1	–	4255	–	3249	33
0.05 g kg^{-1}	Bausch & Lomb	1	–	4120	–	3108	33
0.05 g kg^{-1}	Gilford 202	1	–	4149	–	3096	33
0.06006 g kg^{-1}	Special	1	3644	4232	1420	3130	26[c]
0.06006 g kg^{-1}	NS	NS	3664	4237	1430	3135	32
0.1 g kg^{-1}	Cary	5	–	4298	–	3168	33
0.1 g kg^{-1}	Bausch & Lomb	1	–	4178	–	3139	33
0.1 g kg^{-1}	Bausch & Lomb	1	–	4272	–	3145	33
0.1 g kg^{-1}	Zeiss PMQ-11	1	–	4260	–	3142	33
0.1 g kg^{-1}	Beckman DU	20	–	4278	–	3139	33
0.1 g kg^{-1}	Beckman DB	14	–	4207	–	3124	33
0.1 g kg^{-1}	Beckman DK	5	–	4369	–	3239	33
0.1 g kg^{-1}	Cary	2	–	4272	–	3157	33
0.1 g kg^{-1}	Perkin-Elmer 101	1	–	4284	–	3077	33
0.1 g kg^{-1}	Perkin-Elmer 111	1	–	4481	–	3227	33
0.1 g kg^{-1}	Coleman-Hitachi 124	2	–	4325	–	3116	33
0.1 g kg^{-1}	Hitachi-Perkin-Elmer 139	3	–	4183	–	3183	33
0.1 g kg^{-1}	Perkin-Elmer 202	1	–	4089	–	3001	33
0.1 g kg^{-1}	Perkin-Elmer 350	1	–	4560	–	3295	33
0.1 g kg^{-1}	Perkin-Elmer 402	1	–	4178	–	2913	33
0.1 g kg^{-1}	Bausch & Lomb	1	–	4501	–	3324	33
0.1 g kg^{-1}	Bausch & Lomb	1	–	4413	–	3236	33
0.1 g kg^{-1}	Gilford 202	1	–	4257	–	3136	33
0.05 g l^{-1}[d]	NBS special	1	3659	4249	1423	3151	2[e]
0.05 g l^{-1}[f]	NBS special	1	3653	4248	1424	3151	2[e]
0.05 g l^{-1}[g]	NBS special	1	3649	4245	1424	3154	2[e]
0.05 g l^{-1}[h]	NBS special	1	3642	4244	1425	3159	2[e]
0.02 g kg^{-1}[i]	IMR	1	3602	4192	1411	3136	10[j]
0.04 g kg^{-1}[i]	IMR	1	3616	4209	1413	3140	10[j]
0.06 g kg^{-1}	IMR	1	3622	4227	1415	3144	10[j]
0.08 g kg^{-1}	IMR	1	3644	4245	1417	3148	10[j]
0.10 g kg^{-1}	IMR	1	3659	4263	1419	3152	10[j]
0.05 g l^{-1}[k]	NBS special	1	3647	4251	1426	3156	2[e]
0.05 g l^{-1}[l]	NBS special	1	3643	4246	1425	3160	2[e]
NS[m]	Cary 16	NS	–	–	–	3148	45
NS[m]	Cary 16	NS	–	–	–	3142	46
NS[m]	NS	NS	3655	4263	1430	3171	47

NS: Not specified.
a: water used as reference;   *b*: Average of results from [40] and [41];   *c*: at 21°C; *d*: In H_2SO_4, pH = 1.90;   *e*: At 25°C;   *f*: In H_2SO_4, pH = 1.98;   *g*: In H_2SO_4, pH = 2.20; *h*: In H_2SO_4, pH = 3.00;   *i*: In 0.001 N perchloric acid;   *j*: At 23°C;   *k*: In perchloric acid, pH = 1.99;   *l*: In perchloric acid, pH = 3.08;   *m*: Solvent not specified

Table 4.2: *Mean values for the molar absorptivity of acidic potassium dichromate solution from a collaborative test [12].*
The 'reference' instruments were expected to give more reliable results than the others. The numbers in parentheses are the numbers of instruments in each group.

Concentration (g l^{-1})	λ (nm)	Survey instruments											Reference instruments				
		Beckman DB	Beckman DBG	Beckman DK2 DK2A	Beckman DU	Coleman Hitachi 124	Gilford 240	Gilford 2000, 222	Hitachi-Perkin-Elmer	Zeiss	Other	Mean	Cary	Other	Beckman DU	Beckman ACTA D	Mean
		(11)	(14)	(5)	(7)	(31)	(5)	(5)	(6)	(5)	(37)		(1)	(1)	(1)	(1)	
0.02	240	3471	3722	4016	3736	3972	3677	3780	3795	3648	3913	3839	3780	3825	3677	3736	3751
0.02	257	3972	4148	4413	4222	4413	4075	4133	4295	2810	4325	4251	4251	4251	4133	4163	4207
0.02	300	1809	1912	2089	1986	2133	1868	2030	2015	1633	2045	2000	1912	1956	1912	1912	1927
0.02	350	3015	2957	3221	3163	3324	3104	3104	3163	3001	3133	3148	3133	3163	3089	3825	3295
0.06	240	3467	3530	3844	3722	3810	3687	3599	3726	3668	3702	3702	3731	3731	3672	3658	3697
0.06	257	4035	4055	4339	4207	4305	4163	4060	4271	2863	4227	4197	4231	4231	4192	4129	4197
0.06	300	1863	1927	2054	1951	2054	1912	1868	1986	2069	1966	1971	1893	1927	1888	1834	1888
0.06	350	3001	2922	3143	3118	3207	3089	3020	3133	3099	3064	3099	3109	3118	3123	3035	3099
0.10	240	4119	3492	3813	3651	3689	3654	3580	4075	3930	3695	3651	3727	3719	3619	3672	3683
0.10	257	4025	4128	4281	4192	4166	4148	4066	4457	2854	4178	4178	4248	4231	4236	4175	4225
0.10	300	1886	1915	1942	1936	1992	1903	2483	1986	1862	1971	1953	1953	1930	1921	1871	1918
0.10	350	2927	2960	3145	3071	3127	3063	2989	3168	3124	3033	3057	3130	3118	3118	3071	3110

given separately in Table 4.2 because the wavelengths chosen for the measurements differ slightly from those of other workers.

It has been suggested that one of the isosbestic wavelengths of the $Cr_2O_7^{2-}$ and $HCrO_4^-$ spectra, say 345 nm, could be used for standard absorbance measurements, since there should be no deviation from Beer's Law [10]. However there are small deviations present at 345 nm.

(b)   *Potassium dichromate in alkaline solution*

Alkaline potassium dichromate has not been used extensively in collaborative trials. However, Haupt of the NBS has made detailed studies and recommends that, as potassium dichromate is commercially available in a purer state than potassium chromate, the former should be used to make the solutions - even though the solutions contain effectively only potassium chromate and are identical with solutions of potassium chromate in 0.05 N potassium hydroxide [13]. Fig. 4.3 gives a typical absorption spectrum.

Haupt proposes that the standard solution should be prepared from 0.0300 g of potassium dichromate in 1 litre of 0.05 N aqueous KOH, which will give a solution containing 0.0400 g l^{-1} potassium chromate. The solution was found to be spectroscopically stable above 260 nm over a period of six years. Changes at wavelengths below 260 nm were observed after six months, and so care must be taken if measurements are to be made at short wavelengths. The solutions should be stored in glass bottles - there will be a tendency

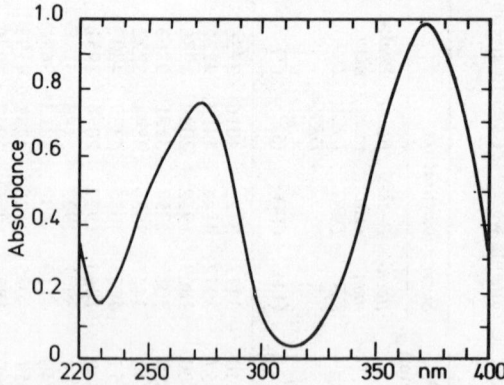

Fig. 4.3   *Absorption spectrum of 1.0×10^{-4} M potassium dichromate in 0.05 N potassium hydroxide measured in a 10 mm cell. Temperature and pH not specified. Redrawn from [3].*

for 'flaking' to occur, but if this is allowed to settle, it should not affect the absorbance values. The solutions obey Beer's Law, and the temperature coefficient at 373 nm is − 0.09 per cent per °C, and at 274 nm is − 0.06 per cent per °C over the temperature range 17-37°C [2]. The results from various measurements and trials are collected in Table 4.3.

(c) *Potassium chromate in disodium hydrogen phosphate solution*
It has been suggested that 0.05 M Na_2HPO_4 could be used in place of KOH to overcome the difficulties of handling strongly alkaline solutions [2, 14]. Deviations from Beer's Law do not exceed 0.10 per cent. Values for the molar absorptivities at two wavelengths are given in Table 4.4.

(d) *Potassium dichromate in neutral solution*
Haupt has examined neutral solutions of potassium dichromate in water [13]. She found that the results were less reproducible than those obtained with alkaline solutions, probably because the pH was not controlled. The solutions were more stable than the alkaline ones and no 'flaking' occurred.

(e) *Comparison of dichromate systems*
Although the stability and linearity of the alkaline solution has been established, the majority of work has been carried out on the more temperamental acidic solutions. The corrosive properties of KOH seem to be disliked by most workers. The acid spectrum has a better arrangement of absorbance maxima and minima, and since the ϵ values are lower than those of the alkaline spectrum, weighing errors are less critical [3]. Because of the effect of pH upon the equilibria under acid conditions, it is essential that stringent standard conditions are laid down for both pH and potassium dichromate concentration.

Further investigations into the use of disodium hydrogen phosphate should be carried out in order to establish it as a more acceptable basic solvent than potassium hydroxide.

4.2.2 *Potassium nitrate*

Potassium nitrate solutions have been widely examined with a view to their use as absorbance standards. In aqueous solutions, potassium nitrate possesses a single absorption band in the UV region at 302 nm

Table 4.3: *Various measurements of the molar absorptivity of potassium dichromate and potassium chromate in 0.05 N potassium hydroxide.*

Concentration (g l⁻¹)	λ (nm)	ε (M⁻¹ cm⁻¹)	λ (nm)	ε (M⁻¹ cm⁻¹)	Instrument type	Number of instruments	Temp. (°C)	Ref. medium	Ref.
$K_2Cr_2O_7$									
0.03	275	3676	375	4815	Various	5	NS	water	13
0.03	273	3800	370	4815	Cary	12	NS	NS	48
0.03	275	3860	375	4830	Beckman	11	NS	water	48
0.03032	273	3790	370	4820	Cary	13	NS	water	49
0.06061	273	3720	370	4800	Cary	13	NS	water	49
0.09089	273	3710	370	4820	Cary	11	NS	water	49
K_2CrO_4									
0.0136	274	3705	373	4830	NBS special	1	25	NS	2
0.272	274	3698	373	4824	NBS special	1	25	NS	2
0.04	273	3690	373	4800	NBS special	1	25	NS	3
0.0408	274	3691	373	4814	NBS special	1	25	NS	2
NS	272	3666	371.5	4820	NS	NS	NS	NS	47
NS	272	3569	371.5	4731	NS	NS	NS	NS	50
NS	273	3668	373	4800	Beckman DU	NS	25	NS	17
NS	272	3330	372	4520	Spectrograph	NS	NS	NS	51
NS	272	3612	372	4719	Spectrograph	NS	NS	NS	52
NS	272.5	3660	371.5	4830	Special	NS	NS	NS	53
NS	—		373	4830	Various	NS	NS	NS	54

NS: Not specified.

Table 4.4: *Molar absorptivities for potassium chromate in 0.05 M disodium hydrogen phosphate at 25°C [2].*

Concentration (M)	pH	Molar absorptivity (M^{-1} cm^{-1})	
		274 nm	373 nm
7×10^{-5}	9.2	3703	4827
14×10^{-5}	9.1	3697	4820
21×10^{-5}	9.1	3692	4813

with a minimum at 262.5 nm (Fig. 4.4). Several collaborative trials have been conducted using potassium nitrate solutions and several attempts made to establish 'accurate' molar absorptivities. However the wide range of results that have been obtained make its use as a standard dubious, as can be seen from the data of Table 4.5.

Several attempts have been made to explain the discrepancies in the results from the collaborative trials, both between the different trials and within them. Potassium nitrate solutions do not obey Beer's Law, there being a negative deviation with increasing concentration [2, 3, 15, 16]. It is thought that ionic interactions are responsible because of the relatively high concentration of solute and the polar characteristics of the N-O bond [2]. Several workers have examined potassium nitrate solutions over a range of concentration and the range over which the absorbance values are linear with concentration seems to vary between instruments [17, 18, 19]. In general, the system appears to be linear when the maximum absorbance values are in the range 0.5-1.5*A*, though Ball noted a 0.3 per cent

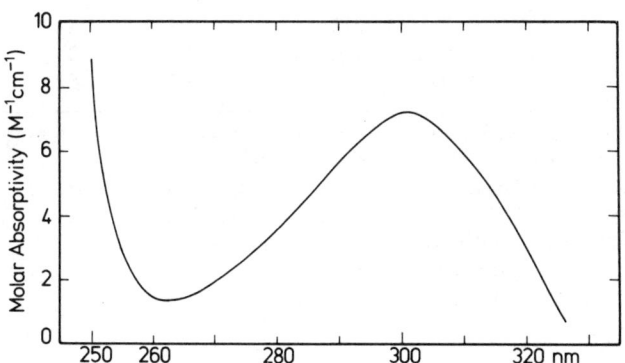

Fig. 4.4 *Absorption spectrum of an aqueous solution of potassium nitrate. Redrawn from [61].*

Table 4.5: *Molar absorptivity values for potassium nitrate solutions.*

Concentration (g l^{-1})	Instrument	Number of instruments	λ (nm)	ϵ (M^{-1} cm^{-1})	T (°C)	Ref.
2.831	NBS special[a]	1	302	7.160	25	2
2.8	NS	NS	302	7.20	NS	1
5.662	NBS special[a]	1	302	7.142	25	2
6.7232	Beckman	1	301.5	7.226	NS	55
7.0-9.6	NBS special[b]	1	302	7.08	25	3
8.492	NBS special[a]	1	302	7.127	25	2
9.9909	Beckman	1	301.5	7.194	NS	55
10.403[c]	Uvispek	NS	301	7.148	NS	21
11.096[d]	Uvispek	NS	301	6.965	NS	21
11.136[e]	Uvispek	NS	301	6.995	NS	21
11.323	NBS special[a]	1	302	7.106	25	2
11.335[f]	Uvispek	NS	301	7.130	NS	21
11.716	Beckman DU	3	302	7.274	NS	22
11.716	Beckman DK1	3	302	7.438	NS	22
11.716	Beckman DK2	18	302	7.196	NS	22
11.716	Unicam SP700	21	302	7.067	NS	22
11.716	Cary 14	6	302	7.067	NS	22
11.716	Perkin-Elmer	3	302	6.981	NS	22
11.716	Perkin-Elmer	11	302	7.188	NS	22
11.716	Hilger	7	302	6.886	NS	22
11.716	Optica	11	302	7.179	NS	22
11.716	Bausch & Lomb	2	302	7.127	NS	22
11.716	NS	8	302	7.119	NS	22
11.716	NPL special	1	301.7±0.2	7.076	25	22
11.716[g]	Uvispek	NS	302	7.071	NS	21
11.716[h]	Uvispek	NS	302	7.067	NS	21
14.14	NBS special[a]	1	302	7.091	25	2
14.0	NS	NS	302	7.11	NS	1
NS	Special	1	302	6.98	NS	56
NS	Special	1	302	7.0	NS	57
NS	Spectrograph	1	303	6.92	NS	58
NS	Spectrograph	1	303	6.92	NS	58
NS	NS	NS	301	7.063	NS	59
NS	Spectrograph	1	301	7.00	NS	60
NS	Beckman DU	NS	301	7.064	25	16

NS: Not specified; *a*: Bandwidth = 1.0 nm; *b*: Bandwidth = 0.6 nm;
c: Recrystallized KNO$_3$ ground, dried at 110°C for 24 h;
d: Recrystallized KNO$_3$; *e*: AR KNO$_3$ dried for 16 h, exposed 6 h and dried 1 h;
f: AR KNO$_3$ dried 16 h; *g*: Stock solution used in Photoelec. Spectr. Grp. trial [22];
h: Ditto from sealed ampoule

difference in ϵ values obtained in this range using 1.0 per cent and 0.67 per cent solutions [19]. Potassium nitrate has a temperature coefficient of −0.1 per cent over the range 17-37°C [2]. Bacterial growth may appear in old solutions, but can be prevented by boiling the water before making the solution, or boiling the solution after it has been prepared [1, 20]. It has been reported that the change in absorbance of a solution is less than 0.1 per cent over a six-month period. The low ϵ value of the peak means that weighing errors in the preparation of the solution are less critical.

Molar absorptivities have been determined for different batches of potassium nitrate that have undergone various purification and drying techniques [21]. These values differed by up to 2 per cent. Experiments to determine the water uptake of the compound gave inconsistent results, and lead to the suggestion that potassium nitrate has variable amounts of water associated with its crystals that cannot all be removed by drying.

The reported variations in ϵ values for potassium nitrate are so large that it should not be used as a standard.

4.2.3 *Pyrene*

Pyrene was used in the PSG collaborative trial of recording instruments in order to reveal 'dynamic' errors, since the spectrum has a number of sharp peaks (Fig. 4.5) [22]. Earlier tests had established that pyrene in iso-octane solution is very stable and showed no spectral changes over a period of one year, although the

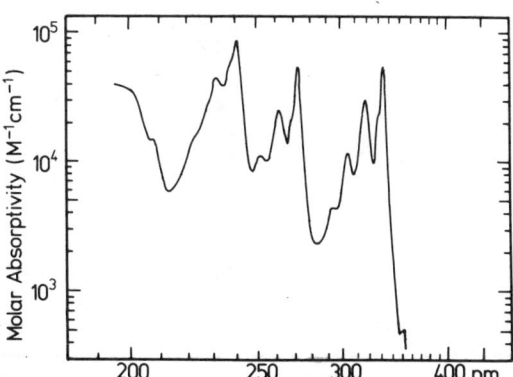

Fig. 4.5 *Absorption spectrum of pyrene in* iso-*octane. Note that the ordinate scale is logarithmic and that the abscissa is linear in frequency. Redrawn from [61].*

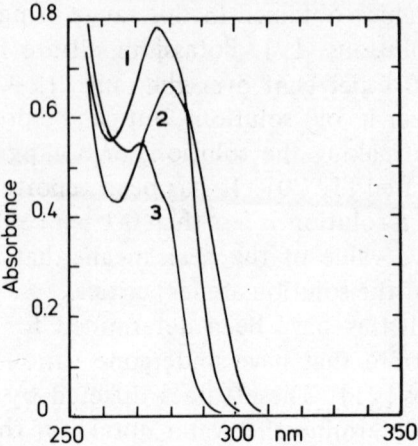

Fig. 4.6   *Absorption spectra of aqueous solutions of potassium hydrogen phthalate measured in a 10 mm cell: (1) 0.122 g l⁻¹ in 0.1 N perchloric acid; (2) 0.103 g l⁻¹ in water; (3) 0.141 g l⁻¹ in 1.0 N potassium hydroxide. Redrawn from [2].*

Table 4.6: *Mean molar absorptivities for pyrene in 180-octane from the Photoelectric Spectrophotometry Group Collaborative Trial, arranged by instrument type [22]*

Instrument	Number of instruments	Molar absorptivity^a (M^{-1} cm^{-1})		
		334 nm	272 nm	240 nm
Unicam SP700^b	1	50 954	49 379	83 935
Unicam SP700	21	52 622	50 861	84 677
Beckman DB	3	54 197	52 066	88 475
Beckman DK1	3	54 660	52 900	92 273
Beckman DK2	18	54 475	52 900	87 456
Cary 11	8	54 845	52 622	87 178
Cary 14	6	55 401	52 714	86 344
Perkin-Elmer 137	11	52 807	51 603	86 344
Perkin-Elmer 4000	3	53 456	51 603	84 769
Hilger Ultrascan	7	52 436	49 935	81 712
Optica CF4	11	53 548	51 417	85 696
Bausch & Lomb 505	2	55 308	52 529	85 788
Manufacturers^c	8	53 085	51 325	86 437
NPL values		334.0±0.1	272.3±0.1	240.6±0.1
		54 475	52 251	86 900
		±556	±556	±556

a: Participants were asked to select slit settings to give an SSW of as near 0.3 nm at 250 nm as possible;   *b:* 15 ampoules were opened and scanned rapidly on a single instrument. The relatively low value is probably the result of the excessive scanning speed, though unfortunately this was not specified.   *c:* These measurements were made by various instrument makers on their own instruments.

Table 4.7: *Molar absorptivities for potassium hydrogen phthalate solutions in 1 per cent perchloric acid, pH = 1.3, at 25°C [2].*

Concentration (g l^{-1})	Min.		Max.	
	λ (nm)	ϵ (M^{-1} cm^{-1})	λ (nm)	ϵ (M^{-1} cm^{-1})
0.034	262.0	917.8	275.5	1293.1
0.142	262.0	916.7	275.5	1290.5

compound does fluoresce [23]. 140 sealed ampoules of 1.1 x 10^{-5} M reagent-grade pyrene in spectroscopic iso-octane were prepared, 15 of these were opened and measured on a Unicam SP700 spectrophotometer and the remainder sent with instructions on measuring technique to the participating members. The results were broken down into several categories; the most representative of these are presented in Table 4.6.

4.2.4 *Potassium hydrogen phthalate*

Potassium hydrogen phthalate is obtainable from the NBS as a high-purity standard for acidimetry. As shown in Fig. 4.6, the spectrum is dependent upon pH, and Burke *et al.* chose 1 per cent aqueous perchloric acid, pH = 1.3, as a solvent [2]. Under these conditions, the solution contains 97 per cent phthalic acid and 3 per cent hydrogen phthalate ions. Their molar absorptivity values at two concentrations are given in Table 4.7.

The temperature coefficient is −0.05 per cent per °C at 275.5 nm, and +0.05 per cent per °C at 262 nm over the range 17-37°C. Weak fluorescence has been observed at 350 nm when exciting at 280 nm but the full effects of this have not yet been estimated.

A comparison of 18 Beckman DU spectrophotometers was carried out by a single operator with 0.01 per cent potassium hydrogen phthalate using a single pair of cells [24]. The results are listed in Table 4.8.

4.2.5 *Picric acid*

The picrate ion has been suggested as an absorbance standard [25]. A 5 x 10^{-3}M solution in aqueous sodium hydroxide was recommended, and at 20°C the maximum at 357.1 nm had an ϵ value of 14 450 M^{-1} cm^{-1}.

Table 4.8: *Molar absorptivites at two wavelengths for potassium hydrogen phthalate in water [24]. The measurements were made on a series of Beckman DU spectrometers by a single operator using a single pair of cells.*

Instrument serial No.	ϵ_{min} 264 nm (M^{-1} cm^{-1})	ϵ_{max} 281 nm (M^{-1} cm^{-1})	$\epsilon_{max}/\epsilon_{min}$
Old instruments:			
1394	878	1301	1.481
2110	878	1311	1.492
2467	872	1299	1.490
New instruments:			
2773	876	1307	1.4912
2764	875	1302	1.4887
2760	877	1304	1.4867
2765	873	1299	1.4877
2770	876	1302	1.4863
2769	877	1306	1.4877
2767	874	1303	1.490
2774	874	1301	1.488
2768	876	1305	1.490
2777	876	1303	1.487
2775	874	1303	1.490
2783	876	1305	1.490
2771	872	1299	1.489
2772	878	1311	1.493
2776	874	1303	1.490
Mean	875	1304	1.489
Relative standard error	0.23%	0.27%	0.19%

4.2.6   *Nicotinic acid*

Nicotinic acid was proposed by Milazzo *et al.* as a suitable absorbance standard for shorter wavelengths [26]. The compound is available in a highly pure state and, if 0.1 N hydrochloric acid is used as solvent, the amino group will be fully protonated. The spectrum is shown in Fig. 4.7, and has a shoulder at 210 nm and a peak at approximately 265 nm. A solution containing 16.399 g l^{-1} in 0.1 N HCl gave a mean absorbance in a 10 mm cell of 0.753Å at 210 nm, with a standard deviation of 0.008 from 70 readings.

Measurements at 210 nm will almost certainly be affected by stray-light, but Milazzo *et al.* claim to have compensated for this by making their measurements in a double series.

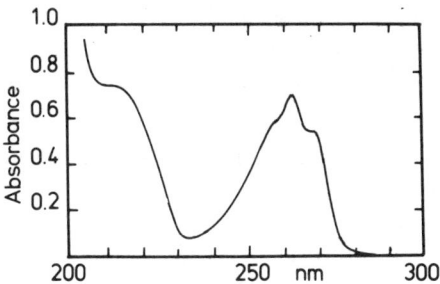

Fig. 4.7 *Absorption spectrum of 0.164 g kg^{-1} nicotinic acid in 0.1 N hydrochloric acid measured in a 10 mm cell [26].*

4.2.7 *Other compounds*

Several compounds were suggested by Hartree as possible solution standards [18]. Two of these, salicylaldehyde in ethanol (λ_{max} = 326 nm) and anthraquinone in ethanol (λ_{max} = 323 nm) were also suggested by Morton as standards for Vitamin A assays [27], ·and ϵ values are given by Morton *et al.* [28, 29] and by Vandenbelt *et al.* [17]. Acridine in ethanol (λ_{max} = 339.5 nm) and caffeine in water (λ_{max} = 272 nm) were also suggested by Hartree, but all four compounds show fluorescence and are not therefore recommended.

4.2.8 *Choice of standards for the 200-400 nm region*

Of the standards reviewed in this section, potassium dichromate has been studied the most and, despite some unsatisfactory properties, potassium dichromate in 0.01 N sulphuric acid seems to be the best standard to use. Great care must be taken to control the pH of the solution. Alkaline solutions have been shown to be stable although 'flaking' may occur. These solutions obey Beer's Law and the absorbance values are less sensitive to pH than the acid solutions. There are not sufficient data on the use of aqueous disodium hydrogen phosphate as a solvent to recommend its use in making a standard. There is no evidence that the mean result from collaborative measurements in a number of laboratories gives a better value for molar absorptivity than a single measurement from a laboratory of excellence. For this reason, the NBS or NPL values for a particular standard are recommended.

The spectrum of potassium hydrogen phthalate is again very pH-dependent and little work has been done on these solutions beyond that reported by the NBS. Potassium nitrate appears to be an ideal standard, with a single broad peak of low ϵ value. However the uncertainty over the ϵ values and the water content of the crystals, and the possibility of bacterial growth make this an unsatisfactory standard. Nictotinic acid is not recommended because most instruments will have stray-light problems at the wavelength of the relevant shoulder. The sharpness of the pyrene bands makes it unsatisfactory because of spectral bandwidth problems and because the exact maxima are difficult to locate. Picric acid has not yet been fully investigated, and the inherent fluorescence of salicylaldehyde, anthraquinone, acridine and caffeine make them unacceptable as standards.

4.3   Standards for the 400-800 nm region

4.3.1   *Thomson's solution*

Thomson has proposed a solution standard that absorbs all through this region (Fig. 4.8) [30]. One litre of the solution contains: 16.67 g $Cr_2 (SO_4)_3 \cdot K_2 SO_4 \cdot 24H_2O$; 33.33 g $CuSO_4 \cdot 5H_2O$; 39.50 g $CoSO_4 (NH_4)_2 SO_4 \cdot 6H_2O$; and 0.120 g $K_2 Cr_2 O_7$. The cobalt ammonium sulphate should be recrystallized, and the other compounds should be of analytical grade. The solution is allowed to 'age' for two months in a glass-stoppered bottle at room temperature. During this time the solution changes from pink to grey in colour, and the pH at 19°C falls from 3.20 to 2.64. The absorption characteristics of an aged solution are given in Table 4.9.

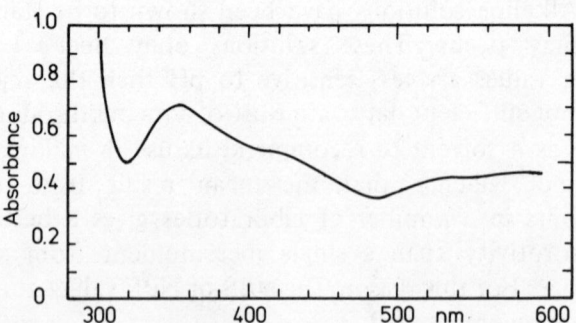

Fig. 4.8   *Absorption spectrum of Thomson's solution made up and aged as described in the text and measured in a 10 mm cell. Redrawn from [1].*

Table 4.9: *Absorbance values for Thomson's solution made up as described in the text and measured in a 10 mm cell [30]. The absorbance values are subject to an uncertainty of ±0.03A.*

λ (nm)	A	λ (nm)	A	λ (nm)	A
420	0.88	490	0.70	620	0.70
430	0.88	500	0.70	640	0.68
440	0.79	520	0.76	660	0.74
450	0.71	540	0.71	680	0.74
460	0.70	560	0.75	700	0.97
470	0.65	580	0.74	720	1.11
480	0.64	600	0.74		

Table 4.10A: *Effects of storage, dilution and contamination upon the absorbance of Thomson's solution prepared as described in the text and measured in a 10 mm cell [30].*

λ (nm)	Absorbance								
	Effect of storage			Addition of equal vol. water, in 20 mm cell		Addition of contaminants			
	0 month	9 month	17 month	Day 1	Day 12	Alcohol	Dirt	Cork	Cotton wool
620	0.72	0.74	0.72	0.74	0.72	0.74	0.70	0.72	0.71
560	0.76	0.75	0.76	0.76	0.74	0.79	0.78	0.76	0.78
490	0.70	0.67	0.69	0.66	0.65	0.66	0.63	0.66	0.65
440	0.79	0.80	0.80	0.78	0.80	0.76	0.80	0.73	0.80

Table 4.10B: *Effects of heating and cooling Thomson's solution. Aliquots of the solution were heated or cooled to the specified temperature and then returned to room temperature for absorbance measurement [30].*

λ (nm)	Absorbance after being at						
	55°C	45°C	35°C	25°C	15°C	5°C	Frozen
620	0.84	0.80	0.75	0.70	0.72	0.74	0.74
560	0.88	0.84	0.82	0.80	0.75	0.76	0.76
490	0.74	0.72	0.71	0.69	0.67	0.71	0.69
440	0.90	0.88	0.84	0.83	0.80	0.80	0.75

The stability of the solution with time and the effects of temperature change and dilution were studied. The results are summarized in Table 4.10. The stability over long periods of time has been confirmed by the NBS [1] and, since temperature changes do affect its performance, it is suggested that the solution is aged and stored at a constant temperature [2]. The temperature coefficient was also measured at 350, 487 and 590 nm; it was found to be greatest at 487 nm where an increase in absorbance from 0.98 to 1.03A on going from 15 to 40°C was observed [1].

Thomson's solution was used in a survey of 132 high-resolution spectrophotometers (i.e. with SSW less than 10 nm) [12]. The solution was sent to the participants in sealed ampoules. No specific measuring instructions were sent but the participants were asked to calibrate their wavelength scales, and check the cleanliness of the cells, their measurements etc. The results are given in Table 4.11.

4.3.2   Cobalt(II) ions

Cobalt(II) as the sulphate or perchlorate has been used as an absorbance standard. Solutions were prepared by dissolving high-purity metal in H_2SO_4-HNO_3 or $HClO_4$-HNO_3 mixtures. The excess nitrate is removed by fuming until the final diluted solution has a pH of 1. The spectrum of the cobalt(II) ion is shown in Fig. 4.9 and has a relatively broad peak of $\lambda_{max} = 512$ nm; ϵ values are given in Table 4.12. The molar absorptivities are temperature-dependent, the temperature coefficient at the maximum being +0.18 per cent per °C over the range 17-37°C.

The higher values in sulphuric acid suggest that a complex is formed with SO_4^{2-} or HSO_4^-, but n.m.r. data show that the $Co(H_2O)_6^{2+}$ octahedral complex is in equilibrium with a $Co(H_2O)_4^{2+}$ tetrahedral complex, which accounts for the solvent effect and the large temperature coefficient [31]. Sets of three solutions of cobalt nitrate in aqueous perchloric acid are available from the Office of Standard

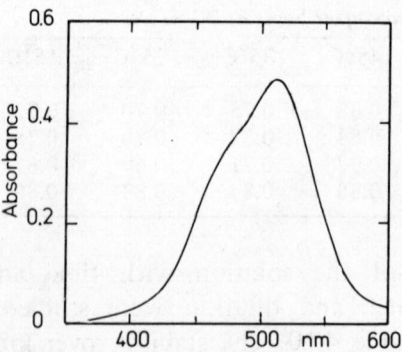

Fig. 4.9   *Absorption spectrum of aquocobalt(II) ion, $Co(H_2O)_6^{2+}$, sulphuric acid, concentration unspecified [2].*

Table 4.11: The absorbance values for Thomson's solution in a 10 mm cell at three concentrations and at three wavelengths determined in a collaborative test [12].

The numbers in parentheses are the number of instruments in each group. The 'reference' instruments were thought to be more reliable.

Relative concentration	λ (nm)	Participating laboratories											Reference instruments				
		Beckman DB	Beckman DBG	Beckman DK2 DK2A	Beckman DU	Coleman-Hitachi 124	Gilford 240	Gilford 2000 & 222	Hitachi-Perkin-Elmer	Zeiss PMQ	Other	Mean	Cary	Other	Beckman DU	Beckman ACTA D	Mean
		(11)	(14)	(5)	(7)	(31)	(5)	(5)	(6)	(5)	(37)		(1)	(1)	(1)	(1)	(1)
0.2	450	0.149	0.147	0.155	0.148	0.153	0.154	0.150	0.151	0.133	0.145	0.149	0.147	0.150	0.148	0.148	0.148
0.2	550	0.136	0.132	0.160	0.134	0.139	0.140	0.134	0.135	0.120	0.131	0.134	0.133	0.132	0.133	0.131	0.132
0.2	650	0.135	0.132	0.160	0.132	0.139	0.138	0.134	0.134	0.123	0.129	0.134	0.133	0.132	0.132	0.131	0.132
0.6	450	0.439	0.445	0.455	0.450	0.460	0.450	0.437	0.459	0.437	0.450	0.449	0.447	0.455	0.450	0.442	0.449
0.6	550	0.405	0.403	0.470	0.409	0.418	0.411	0.398	0.415	0.396	0.408	0.409	0.407	0.407	0.401	0.403	0.405
0.6	650	0.406	0.403	0.464	0.405	0.416	0.411	0.400	0.413	0.400	0.403	0.407	0.404	0.408	0.405	0.401	0.405
1.0	450	0.735	0.749	0.758	0.747	0.741	0.760	0.742	0.741	0.768	0.750	0.750	0.750	0.759	0.752	0.743	0.751
1.0	550	0.678	0.678	0.772	0.681	0.672	0.691	0.678	0.671	0.682	0.678	0.682	0.680	0.681	0.671	0.670	0.676
1.0	650	0.681	0.674	0.767	0.670	0.670	0.687	0.676	0.673	0.681	0.670	0.678	0.673	0.683	0.676	0.677	0.677

Table 4.12: *Molar absorptivities for the cobalt(II) ion in 0.1 M sulphuric acid and 0.1 M perchloric acid at 25°C [2, 3]*

Concentration of $C_0(II)$ (gl^{-1})	Molar absorptivity at 512 nm (M^{-1} cm^{-1})	
	sulphate	perchlorate
2.0	4.87	4.82
2.4	4.88	4.80
12.0	4.87	4.79

Reference Materials, Washington [32]. Certificates giving absorbance values at 302, 395, 512 and 678 nm measured at 25°C are provided.

4.3.3   *Cobalt ammonium sulphate*

This compound appears to be more widely used as a standard than cobalt sulphate and has been utilized in several surveys [33, 34]. An aqueous acid solution obeys Beer's Law over a considerable absorbance range, and so by varying the concentration a wide photometric range can be covered [35]. A typical absorption spectrum is shown in Fig 4.10. The absorbance of the solution at wavelengths up to 680 nm increases with temperature [36].

The data of Table 4.13 were obtained from a solution of 14.481 g

Table 4.13: *Molar absorptivities for cobalt ammonium sulphate $CoSO_4(NH_4)_2SO_4 \cdot 6H_2O$ in 1 per cent sulphuric acid at 25°C [25, 35]*

λ (nm)	ϵ (M^{-1} cm^{-1})	λ (nm)	ϵ (M^{-1} cm^{-1})	λ (nm)	ϵ (M^{-1} cm^{-1})
350	0.1037	520	4.6100	690	0.1801
360	0.1092	530	3.9637	700*	0.1474
370	0.1365	540	3.0381	710	0.1255
380	0.1774	550*	2.1154	720	0.1037
390	0.2402	560	1.3539	730	0.0873
400*	0.3412	570	0.8408	740	0.0819
410	0.4586	580	0.5651	750*	0.0764
420	0.6114	590	0.4313		
430	0.9281	600*	0.3739	Hg 404.7*	0.3931[+]
440	1.4249	610	0.3384	Hg 435.8*	1.1929[+]
450*	2.1101	620	0.3139	Hg 491.6*	4.0860[+]
460	2.8145	630	0.3057	He 501.6*	4.5342[+]
470	3.3113	640	0.3003	Hg 546.1*	2.4592[+]
480	3.6821	650*	0.2866	Hg 578.0*	0.5977[+]
490	4.0179	660	0.2647	He 578.6*	0.4558[+]
500*	4.4623	670	0.2374	He 667.8*	0.2429[+]
510	4.7544	680	0.2074		

* These values are considered by the authors to be the most reliable.

+ These values have been adopted by the IUPAC [32].

Table 4.14: *Molar absorptivities of cobalt ammonium sulphate, $CoSO_4(NH_4)_2SO_4 \cdot 6H_2O$ in 1 per cent sulphuric acid at two concentrations [42].*

λ (nm)	Molar absorptivity (M^{-1} cm^{-1})	
	0.0367 M	0.0735 M
400	0.3270	0.3537
450	2.0436	2.0680
500	4.3597	4.4354
510	4.7411	4.7076
550	2.0708	1.9864
600	0.1090	0.2993

$CoSO_4(NH_4)_2 \cdot 6H_2O$ (material of purity greater than 99.9 per cent is available) in 1 litre of 1 per cent sulphuric acid (specific gravity = 1.835) in water [25, 35]. The values marked with an asterisk are cited by the IUPAC Commission on Physicochemical Measurements and Standards [32], although the concentration of their solution is greater, namely 14.821 $g\,l^{-1}$. Rand has measured molar absorptivities at two concentrations of the salt, and these are given in Table 4.14.

A survey has been carried out by Vanderlinde *et al.* on the performance of cobalt ammonium sulphate solutions measured with various spectrometers, divided into two groups of SSW greater and less than 10 nm [33]. Two solutions, of concentration 0.500 ± 0.0001 M and 0.1002 ± 0.0002 M cobalt ammonium sulphate, were supplied to each participant together with a standard procedure for measurement. The results are given in Tables 4.15 and 4.16.

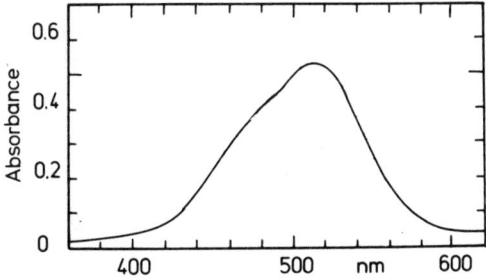

Fig. 4.10 *Absorption spectrum of cobalt ammonium sulphate in 1 per cent sulphuric acid, concentration unspecified [42].*

Table 4.15: *Molar absorptivity values at 510 nm for cobalt ammonium sulphate in 1 per cent sulphuric acid, measured on narrow band (SSW less than 10 nm) instruments[33].*

Instrument	No. of instruments	Molar absorptivity	
		0.0500 M	0.1002 M
REFERENCE INSTRUMENTS			
Cary	4	4.90±0.04	4.87±0.04
Bausch & Lomb Spectronic 100	2	4.88	4.91
Bausch & Lomb Spectronic 500	1	4.88	4.92
Bausch & Lomb Spectronic 600	1	4.80	4.89
All reference instruments	8	4.88±0.04	4.89±0.03
SURVEY INSTRUMENTS			
Beckman DB, DB-G	15	4.92±0.24	4.91±0.13
Gilford 300	15	4.76±0.18	4.76±0.19
Beckman DU, DU-2	11	4.92±0.20	4.90±0.15
Bausch & Lomb Spectronic 100	4	4.86	4.89
Beckman B	4	4.76	4.80
Hitachi-Perkin-Elmer 139	4	4.74	4.81
Cary 14	1	4.90	4.94
Gilford 2000	1	4.72	4.90
Perkin-Elmer 202	1	4.60	4.54
Turner 330	1	4.60	4.54
Zeiss PM Q-11	1	5.00	4.99
ALL SURVEY INSTRUMENTS	58	4.84±0.22	4.85±0.17

4.3.4   Nickel ions

Nickel(II) ions as the sulphate and perchlorate have been studied with a view to their use as standards [2]. The solutions were prepared in the same way as the cobalt solutions of Section 4.3.2, and a typical spectrum is given in Fig. 4.11. The NBS data for the

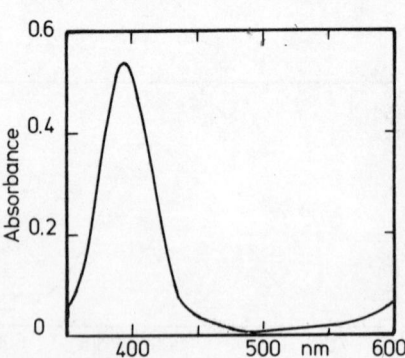

Fig. 4.11   *Absorption spectrum of a quonickel(II) ion, $Ni(H_2O)_6{}^{2+}$, in sulphuric acid, concentration unspecified [2].*

Table 4.16: *Molar absorptivity values at 510 nm for cobalt ammonium sulphate in 1 per cent sulphuric acid, measured on broad band (SSW greater than 10 nm) instruments [33].*

Instrument	Cell type	No. of instruments	Molar absorptivity (M^{-1} cm^{-1})	
			0.0500 M	0.1002 M
Bausch & Lomb Spectronic 20	NS	75	4.56 ± 0.24	4.62 ± 0.25
Coleman Universal	13 mm Rect.	2	4.24	4.27
Coleman Junior 35 nm SSW	13 mm Rect.	5	4.40	4.42
Coleman Junior 6A, 6B & 6C 25 nm SSW	10 mm Tubular	34	3.82 ± 0.34*	3.81 ± 0.39*
	12 mm Tubular	58	3.94 ± 0.22*	3.85 ± 0.22*
	19 mm Tubular	67	3.86 ± 0.26*	3.67 ± 0.27*
Coleman Junior 6D 20 nm SSW	10 mm Tubular	33	4.04 ± 0.44*	3.84 ± 0.44*
	12 mm Tubular	74	4.28 ± 0.20*	4.15 ± 0.18
	19 mm Tubular	51	4.12 ± 0.16*	3.99 ± 0.19
Dow Diagnotest	NS	6	4.44 ± 0.04	4.48 ± 0.16
Eppendorf	NS	1	2.62	2.51
Eskalab	NS	2	4.90	4.95
Fisher Clinical Electrophotometer	NS	2	3.52	3.49
Leitz M	13 mm Rect.	78	3.60 ± 0.36	3.55 ± 0.33
Lumetron 14	18 mm Tubular	12	2.68 ± 0.20	2.52 ± 0.13
Klett	12.5 mm Tubular	48	3.28 ± 0.18	3.23 ± 0.14
Serometer 360	13 mm Tubular	3	2.92	2.81

NS: Not specified; SSW: Spectral slitwidth; *These means were recalculated omitting values falling outside 3 × standard deviation.

Table 4.17: *Molar absorptivities for the nickel(II) ion in 0.1 M sulphuric acid and 0.1 M perchloric acid at 25°C [2,3]*

Concentration of Ni(II) ($g l^{-1}$)	Molar absorptivity ($M^{-1} cm^{-1}$)	
	sulphate at 394 nm	perchlorate at 391 nm
2.3	5.17	5.09
11.5	5.15	5.07
unspecified	5.17	5.11

maximum molar absorptivity are given in Table 4.17.

Nickel (II) nitrate solutions containing 6.8 g l^{-1} of nickel at pH = 2.7 have been measured by the NBS at 394 nm [1]. Measurements were made with an SSW of 1.5 nm and a value of ϵ_{394} = 5.09 at 25.0 ± 0.1°C is given. The temperature coefficient is +0.14 per cent per °C over the temperature range 17-37°C. As with the cobalt salts, the values for the sulphate are greater than for the perchlorate: the explanation is probably the same.

The Office of Standard Reference Materials, Washington, provide a set of three solutions of nickel(II) nitrate in aqueous perchloric acid, with a certificate showing their absorbance at 302, 395, 512 and 678 nm at 25°C [32].

4.3.5   *Copper(II) ion*

Acidic solutions of copper(II) sulphate are stable and have been proposed as standards over the range 350-750 nm. A typical spectrum is shown in Fig. 4.12. The solution obeys Beer's Law over a wide

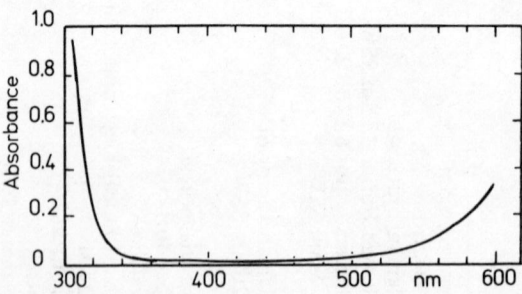

Fig. 4.12 *Absorption spectrum of copper(II) sulphate in aqueous solution, concentration unspecified [1].*

Table 4.18: *Molar absorptivities for copper sulphate (CuSO₄·5H₂O) in 1 per cent sulphuric acid at 25°C [25, 35].*

λ (nm)	ε (M⁻¹ cm⁻¹)	λ (nm)	ε (M⁻¹ cm⁻¹)	λ (nm)	ε (M⁻¹ cm⁻¹)
350	0.1124	520	0.0687	690	5.7306
360	0.0786	530	0.0986	700*	6.5796
370	0.0574	540	0.1386	710	7.3909
380	0.0437	550*	0.1935	720	8.1903
390	0.0349	560	0.2697	730	9.5896
400*	0.0287	570	0.3646	740	9.5896
410	0.0237	580	0.4870	750*	10.2000
420	0.0200	590	0.6467		
430	0.0175	600*	0.8490	Hg 404.7*	0.0262^+
440	0.0150	610	1.1048	Hg 435.8*	0.0162^+
450*	0.0137	620	1.4048	Hg 491.6*	0.0237^+
460	0.0137	630	1.7853	He 501.6*	0.0350^+
470	0.0150	640	2.2475	Hg 546.1*	0.1685^+
480	0.0175	650*	2.7964	Hg 587.0*	0.4594^+
490	0.0225	660	3.4214	He 587.6*	0.6080^+
500*	0.0324	670	4.1448	He 667.8*	3.9829^+
510	0.0474	680	4.8944		

* These values are considered by the authors to be the most reliable.

+ ｜These values have been adopted by IUPAC [32].

concentration range and a wide photometric range can be covered. The low ε values means that weighing errors are not important. Above 520 nm the absorbance of the solutions increases with temperature, the coefficient increasing with wavelength. Below 520 nm the changes are small over the range 25-40°C[36].

The standard solution is prepared by dissolving 20.000 g of $CuSO_4 \cdot 5H_2O$ (purity greater than 99.9 per cent) in 1 litre of 1 per cent sulphuric acid (specific gravity = 1.835) in water. Values obtained with this solution are given in Table 4.18. Those marked with an asterisk are cited by the IUPAC Commission on Physicochemical Measurements and Standards [32].

4.3.6 *Iron-dipyridyl complex*

Hartree suggested that the iron-a, a'-dipyridyl complex ion, $Fe(C_5H_4N)_2{}^{2+}$, could be employed as an absorbance standard [18]. The complex has a maximum at 520 nm, and is formed by mixing equal volumes of 0.05 M ferrous ammonium sulphate and 0.3 M a, a'-dipyridyl in 0.6 N H_2SO_4. The use of a complex of this type, whose absorbance is strongly dependent upon an equilibrium reaction, is not recommended.

4.3.7   *Choice of standards for the 400-800 nm region*

The best standard in this group would appear to be acidic cobalt ammonium sulphate. The solution is readily prepared from commercially available high-purity material and the spectral properties have been thoroughly evaluated.

Thomson's solution is stable once it has been 'aged'. The spectrum is relatively smooth and so it can be used throughout this spectral range but, because it is a mixture, no ϵ values can be calculated. Accordingly, the preferred application is as a linearity or transfer standard.

Solutions of cobalt(II) and nickel(II) ions have been examined by the NBS, both as the sulphate and the perchlorate. The preparation of the solutions is not easy and so their use is not recommended. The solutions can be purchased from the NBS. Acidic copper sulphate has no absorption maximum in this region which means that it is not wholly satisfactory as an absorbance standard. Since it follows Beer's Law over a wide concentration range, it is a useful linearity standard.

4.4   Standards for the entire 200-800 nm region

4.4.1   *The NBS 'Composite' solution*

The NBS Composite solution was formulated to extend the working range of Thomson's solution below 300 nm (Fig. 4.13) [1]. This was achieved by using the chromium(III) and cobalt(II) components of the original solution together with *p*-nitrophenol in sulphuric acid at pH = 1. The solution does not appear to fluoresce and has several

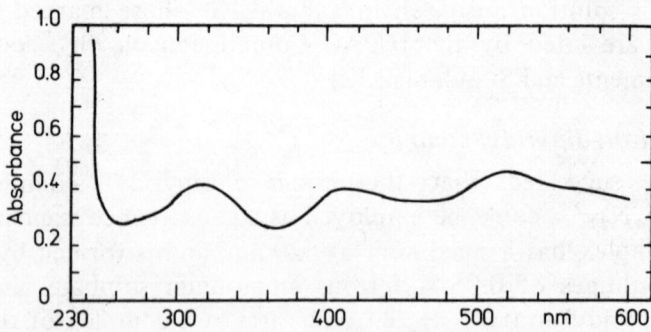

Fig. 4.13 *Absorption spectrum of NBS Composite solution, pathlength unspecified [1].*

Table 4.19: *Stability of the NBS Composite solution [1]*

λ (nm)	Absorbance in 10 mm cell	
	After ageing	after 15 weeks
525	0.562	0.564
470	0.397	0.398
415	0.483	0.484
365	0.252	0.253
320	0.493	0.493
260	0.256	0.258

broad maxima and minima over the range 250-600 nm [2]. The solution must be 'aged' for two months at room temperature before use and, as it is liable to suffer an irreversible absorbance change with temperature, it must be stored at constant temperature. The temperature coefficients of the maxima at 320, 415 and 525 nm were determined and it was found that the 525 nm peak was most sensitive, the absorbance increasing from 0.98 to 1.025 on going from 15 to 40°C [1]. Table 4.19 shows that, once aged, the solution is stable for at least three and a half months.

4.4.2 Organic dyes

The NBS have examined a number of dyes suitable for use as standards [1]. The dyes studied so far are Neolan Black, Cibalan Black, Alizarin Light Grey, Luxol Fast Black L and Nigrosine. As Fig. 4.14 shows, the dyes have maxima in two regions: 220-250 nm and 550-600 nm. They have high ϵ values and the concentrations required would be in the range 0.04-0.06 g l^{-1}. The first three dyes have been studied in depth. Their solutions are insensitive to pH in the range 2-9, and boiling produces no significant spectral changes.

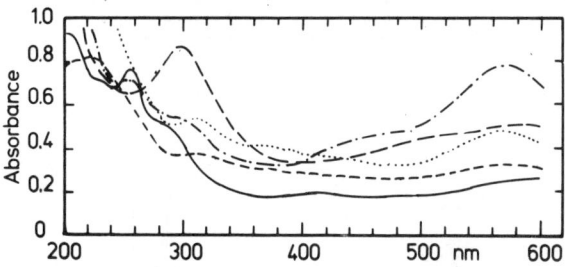

Fig. 4.14 *Absorption spectra of 'black' organic dyes in water, concentrations unspecified: Alizarin Light Grey (———); Cibalan black (– – –); Neolan Black (–·–): Nigrosine (- - - -); Luxol Fast Black (. . . . .) [1].*

They are relatively stable and when exposed to fluorescent light the absorbance changes were less than 1 per cent over two months. The temperature coefficients for Cibalan Black in the range 15-40°C measured at 315, 400 and 575 nm did not exceed 0.1 per cent per °C. Similar results were obtained for Neolan Black and Alizarin Light Grey.

The main problem with these dyes is that commercial samples contain up to 40 per cent impurity. At present, difficulties are being found in purifying dyes, although some progress has been made with Alizarin Light Grey [3].

4.4.3   *Green food colourings*

It has been proposed that French's Green Food Colouring is suitable as a linearity standard [37]. The colouring is a mixture of FDC Yellow No. 5 (CI Food Yellow No. 4) and FDC Blue No. 1 (CI Food Blue No. 2) in a ratio of 5:1 at a total concentration of 25 g l^{-1} in aqueous propylene glycol. 0.10 ml of this stock solution is diluted to 200 ml with water, and the resulting solution has been found to be stable for 6 months when stored in a brown glass bottle at room temperature. There is no apparent change in absorbance on varying the pH between 2.5 and 10, or on varying the temperature between 4° and 56°C. Beer's Law was found to be obeyed when the solution was diluted eight-fold. The solution has been employed in twelve American laboratories over a period of twelve months, using four different batches of colouring, all of which gave essentially the same results. An absorption spectrum is given in Fig. 4.15 and the absorbance values are listed in Table 4.20.

French's colouring is not available in Britain, but Langdale's Sap Green, consisting of Green S and Tartrazine, gives similar results [38]. 5.0 ml of the colouring is diluted to 500 ml with water, and 20 ml of this solution is diluted to 100 ml with water. The absorbance

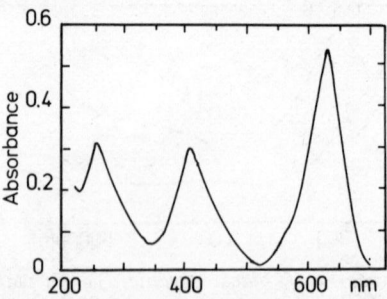

Fig. 4.15   *Absorption spectrum of French's Green Food Colouring, 0.05 per cent of commercial solution in water [37].*

Table 4.20: *Deviations from Beer's Law upon dilution of a solution of French's Green Food Colouring [37].*

Relative concentration	Absorbance in 10 mm cell		
	257 nm	410 nm	630 nm
1.000	0.545	0.979	0.544
0.500	0.283	0.492	0.278
0.250	0.148	0.252	0.140
0.125	0.072	0.124	0.070

was found to be independent of pH over the range 2-6.5, and the values measured at 238, 425 and 634 nm were constant over at least 30 days. A further survey conducted by Burgess used a more concentrated solution made by diluting 5 ml of the Langdale's Sap Green stock solution to 2 litres with water. The results are given in Table 4.21. The solution remained stable for over one year and the only deterioration noted was a slight bacterial growth that did not affect the absorbance values.

These dye solutions can only be used as linearity checks or transfer standards because of the difficulty in obtaining the pure components. They have the advantage of being safe, inexpensive, very stable and easy to use.

4.4.4   *Choice of a standard for the 200-800 nm region*

There is no absorbance standard that will satisfactorily cover the whole UV-visible range. The NBS Composite solution has satisfactory properties, but lack of precise details of its preparation means that it cannot be used as a standard. The dye solutions and food colourings look promising but cannot be used until the pure compounds are readily available and further studies are carried out.

4.5   Commercially available standards

There are now several sources of solutions of certified absorbance that can be used as standards in the UV and visible regions. A selection of these is given below.

4.5.1   *SRM 931: Cobalt and nickel perchlorates*

Standard Reference Material 931,
The Office of Standard Reference Materials,
Washington, D.C. 20234, USA.

Table 4.21: *Typical results with Langdale Sap Green dye solution: effects of storage and light* [15]
This compares measurements made on four concentrations of a stock solution in water measured at three wavelengths in different laboratories at various time intervals.

Site, instrument, storage time & conditions	Absorbance at 239 nm				Absorbance at 424 nm				Absorbance at 634 nm			
	100%	50%	25%	12.5%	100%	50%	25%	12.5%	100%	50%	25%	12.5%
Laboratory A; Perkin-Elmer 552; Initial, Sample 1	2.32	1.159	0.566	0.287	1.167	0.575	0.292	0.147	1.033	0.518	0.254	0.125
Laboratory A; Perkin-Elmer 552; Initial, Sample 2	2.32	1.164	0.582	0.293	1.170	0.589	0.295	0.147	1.042	0.526	0.260	0.131
Laboratory A; Perkin-Elmer 200; 2 months in dark	2.32	1.183	0.593	0.304	1.166	0.585	0.295	0.148	1.057	0.531	0.253	0.134
Laboratory A; Perkin-Elmer 200; 2 months in light	2.32	1.171	0.586	0.301	1.169	0.591	0.296	0.150	†	0.529	0.264	0.132
Laboratory B; Pye Unicam SP8-200; 2 weeks in light	2.318	1.172	0.594	0.296	1.167	0.590	0.297	0.148	1.041	0.524	0.265	0.132
Laboratory C; Hilger Uvispek Mk 9; 1 month in light	(≈2.25)*	1.175	0.594	0.299	(≈1.17)*	0.584	0.298	0.149	†	0.513	0.262	0.132
Laboratory D: Varian Superscan; 1 month in light	2.310	1.164	0.587	0.295	1.167	0.586	0.294	0.148	1.044	0.526	0.263	0.133
Mean	2.318	1.170	0.586	0.296	1.168	0.586	0.295	0.148	1.043	0.524	0.261	0.131
Confidence Limits at 95% level	0.004	0.007	0.009	0.006	0.002	0.005	0.002	0.001	0.011	0.006	0.006	0.003

* bracketed values excluded from analysis
† measurements not made.

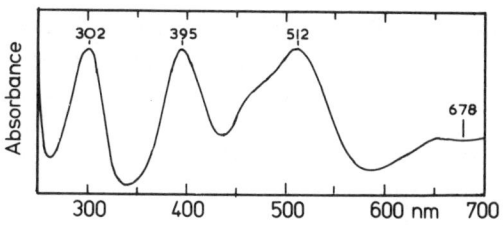

Fig. 4.16   *Absorption spectrum of NBS Standard Reference Material 931. A set of three solutions is supplied, of maximum absorbance approximately 0.3, 0.6 and 0.9 [2].*

This was the first absorbance standard to be issued by the NBS [2]. Solutions of different concentrations and a blank of 0.1 N perchloric acid are supplied in sealed ampoules, together with a certificate of absorbance at four wavelengths. The solutions are prepared by dissolving high-purity cobalt and nickel in a mixture of nitric and perchloric acids. The weights are chosen so that the maxima are of approximately the same height. A typical spectrum is given in Fig. 4.16. The concentration of nitrate ion is reduced to that of the metallic ions by fuming, giving a solution of pH = 1.

The absorbance is given at 302, 395, 512 and 678 nm. The SSW is critical and, to obtain readings comparable to those quoted, it should not exceed 1.0, 1.7, 2.0 and 6.5 nm respectively. The uncertainties in the values are given at the 95 per cent confidence level and include a systematic error of 0.5 per cent. The absorbance is certified at 25°C but the corresponding value at any temperature between 17 and 27°C can be calculated from the following formula:

$$A_t = A_{25} + (t - 25).C_A$$

where A_t and A_{25} are the absorbances at the temperature of observation and 25°C, and C_A is the temperature coefficient, namely −0.0014 at 302 nm; +0.0014 at 395°C; +0.0018 at 512 nm and +0.0014 at 678 nm.

4.5.2   *Harleco standards for the visible region*

AHS (United Kingdom),
Station Road,
Didcot, Berkshire, England.

Harleco supply solutions of nickel(II) sulphate (λ_{max} = 670 nm), cobalt ammonium sulphate (λ_{max} = 512 nm) and chromium(III) perchlorate (λ_{max} = 408 nm) at four or five different concentrations, together with a blank. They are supplied either in ampoules or re-usable 12 mm tubular cells.

4.5.3   *Starna potassium dichromate*

Starna Ltd,
High Road,
Ilford, Essex, England.

Potassium dichromate standards are supplied in six sealed 10 mm cells with nominal absorbances of 0, 0.2, 0.4, 0.6, 0.8 and 1.2*A* at 350 nm. The temperature coefficient is less than 0.1 per cent per °C over the range 20-30°C and the samples are claimed to be stable indefinitely. The manufacturers recommend that these cells should be used as secondary standards only.

4.5.4   *Holnicob visible absorbance standards*

Laboratoires Biotrol,
Rue du Foin, 75140 Paris, France.

Solutions of 0.1, 0.2, and 0.4 M nickel(II) nitrate, 0.025, 0.05, 0.1 and 0.2 M cobalt(II) nitrate, nitric acid blanks and 5 per cent holmium nitrate (for wavelength calibration) are supplied in sealed ampoules. A certificate of absorbance values is supplied with each set. Knowles has examined these sets and found the linearity to be good, indicating careful preparation of the solutions [39]. No mention of temperature control is made, although both nickel and cobalt molar absorptivities are temperature-dependent. Each set can be used only once, and the system is therefore expensive to use routinely.

4.5.5   *SRM 935: Solid potassium dichromate*

The Office of Standard Reference Materials - see Section 4.5.1. SRM 935 is crystalline potassium dichromate of certified purity which is supplied with extensive details of the preparation of solutions, measurement of absorbance and the expected ϵ values. It is described in detail by Burke and Mavrodineanu [4].

4.6   **Conclusions**

The majority of solutions mentioned in this chapter can be quickly prepared from readily available compounds. Most of these solutions are stable over long periods and so a reasonable quantity can be prepared at one time. It is therefore recommended that solution standards are prepared in the laboratory, using the purest available compounds or SRM 935. The commercially available solutions are expensive and offer no particular advantage, unless time or the necessary expertise are at a premium.

Procedures for the preparation and measurement of a dichromate standard are given in Chapter 10.

References

1 Menis, O. and Schultz, J.I. (1970), NBS Technical Note 544.
2 Burke, R.W., Deardorff, E.R. and Menis, O. (1972), *J. Res. NBS,* **76A**, 469.
3 Menis, O. and Schultz, J.I. (1971), NBS Technical Note 584.
4 Burke, R.W. and Mavrodineanu, R. (1977), NBS Special Publication No. 260-54.
5 Ringbom, A. (1963), *Complexation in Analytical Chemistry,* Wiley, New York.
6 Neuss, J.D. and Rieman, W. (1934), *J. Amer, Chem. Soc.,* **56**, 2238.
7 Howard, J.R., Nair, V.S.K. and Nancollas, G.H. (1958), *Trans. Faraday Soc.,* **54**, 1034.
8 Tong, J. and King, E.L. (1953), *J. Amer. Chem. Soc.,* **75**. 6180.
9 Davis, W.G. and Prue, J.E. (1955), *Trans. Faraday Soc.,* **51**, 1045.
10 Burke, R.W. and Mavrodineanu, R. (1976), *J. Res. Nat. Bur. Std.,* **80A**, 631.
11 West, M.A. and Kemp, D.R. (1976), *Int. Lab.,* May/June, 27.
12 Beeler, M.F. (1974), *Amer. J. Clin. Path.,* **61**, 789.
13 Haupt, G.W. (1952), *J. Opt. Soc. Amer.,* **42**, 442.
14 Johnson, E.A. (1967), *Photoelec. Spec. Grp. Bull.,* **17**, 505.
15 Burgess, C., unpublished observations.
16 Von Halband, H. and Eisenbrand, J. (1928), *Z. Phys. Chem.,* **112**, 620.
17 Vandebelt, J.M., Forsyth, J. and Garret, A. (1945), *Ind. Eng. Chem. Anal., (Ed.)* **17**, 235.
18 Hartree, E.F. (1950), *Photoelec. Spec. Grp. Bull.,* **2**, 32.
19 Ball, S. (1948), PhD Thesis, University of Liverpool.
20 Edisbury, J.R. (1966), *Practical Hints on Absorption Spectroscopy,* Hilger & Watts, London.
21 Everett, A.J., unpublished observations.
22 Anon. (1965), *Photoelec. Spec. Grp. Bull.,* **16**, 443.
23 Kendall, G., unpublished observations.
24 Goldring, L.S., Hawes, R.C., Hare, G.H., Beckman, A.O. and Stickney, M.E. (1953), *Anal. Chem.,* **25**, 869.
25 Kortüm, G. (1955), *Kolorimetrie - Photometrie und Specktrometrie,* 3rd Edn, Springer-Verlag, Berlin.
26 Milazzo, G., Palumbo-Doretti, S.C.M. and Violante, N. (1977), *Anal. Chem.,* **49**, 711.

27 Morton, R.A. (1942), *Absorption Spectra,* 2nd Edn, Hilger, London.
28 Morton, R.A. and Stubbs, A.L. (1940), *J. Chem. Soc.,* 1347.
29 Morton, R.A. and Earlam, W.T. (1941), *J. Chem. Soc.,* 159.
30 Thomson, L.C. (1946), *Trans. Faraday Soc.,* **42**, 663.
31 Swift, T.J. and Connick, R.E. (1962), *J. Chem. Phys.,* **37**, 307.
32 Herington, E.F.G. and Milazzo, G. (1976), *Recommended reference materials for the realisation of physicochemical properties,* IUPAC Commission on physicochemical measurements and standards.
33 Vanderlinde, R.E., Richards, A.H. and Kowalski, P. (1975), *Clin. Chem. Acta.,* **61**, 39.
34 Rand, R.N. (1973), NBS Special Publication No. 378, 125.
35 Gibson, K.S. (1949), NBS Circular 484.
36 Davies, R. and Gibson, K.S. (1931), NBS Miscellaneous Publication M114.
37 Frings, C.S., Muscat, V.I. and Waldrop, N.T. (1976), *Clin. Chem.,* **22**, 101.
38 Burgess, C. (1977), *UV Spec. Grp. Bull.,* **5**, 77.
39 Knowles, A. (1977), *UV Spec. Grp. Bull.,* **5**, 94.
40 Gridgeman, N.T. (1951), *Photoelec. Spec. Grp. Bull.,* **4**, 67.
41 Ketelaar, J.A.A., Fahrenfort, J., Haas, C. and Brinkman, G.A. (1955), *Photoelec. Spec. Grp. Bull.,* 8, 176.
42 Rand, R.N. (1969), *Clin. Chem.,* **15**, 839.
43 Bouché, R. and Molle, L. (1975), *J. Pharm. Belg.,* **30**, 578.
44 Vandenbelt, J.M. (1960), *J. Opt. Soc. Amer.,* **50**, 24.
45 Wylie, E., unpublished observations, quoted in Ref. [14].
46 Inman, S.R., unpublished observations, quoted in Ref. [14].
47 Morton, R.A. (1951), *Photoelec. Spec. Grp. Bull.,* **4**, 65.
48 Brode, W.R., Gould, J.H. Whitney, J.E. and Wyman, C.M. (1953), *J. Opt. Soc. Amer.,* **43**, 862.
49 Vandenbelt, J.M. and Spurlock, C.H. (1952), *J. Opt. Soc., Amer.,* **43**, 862.
50 Hogness, T.R. and Zscheile, F.P. (1947), unpublished observations.
51 Rössler, G. (1926), *Chem. Ber.,* **59**, 2606.
52 Morton, R.A., unpublished observation.
53 Hogness, T.R., Zscheile, F.P. and Sidwell, A.E. (1937), *J. Phys. Chem.,* **41**. 379.
54 *Collaborative Spectrophotometric Study* (1940), U.S. Pharmacopoeia.
55 Edisbury, J.R. (1949), *Photoelec. Spec. Grp. Bull.,* **1**, 10.
56 Von Halban, H. and Ebert, L. (1924), *Z. Phys. Chem.,* **112**, 329.
57 Scheibe, G. (1926), *Chem. Ber.,* **59**, 2606.
58 Baly, E.C.C., Morton, R.A. and Riding, R.W. (1927), *Proc. Roy. Soc.,* **113A**, 709.
59 Ley, H. and Volbert, F. (1927), *Z. Phys. Chem.,* **130**, 308.
60 Ewing, D.T., Vandenbelt, J.M., Emmett, A.D. and Bird, O.D. (1940), *Ind. Eng. Chem. Anal. Ed.,* **12**, 639.
61 Perkampus, H., Sandeman, I. and Timmons, C.J. (Eds) (1971), *UV Atlas of Organic Compounds,* Verlag Chemie.

5 Solid absorbance standards

5.1 Introduction

Many methods of checking the absorbance accuracy of spectro-photometers that do not use solutions have been developed over the years. It is the purpose of this chapter to highlight the major papers dealing with these methods and, in particular, to discuss those methods most suited to routine laboratory use.

It is relevant at this point to consider what properties an absorbance standard should possess. Ideally it should:

(a) be convenient to use and should simulate the normal use of the instrument;

(b) have an absorbance independent of wavelength setting;

(c) be unaffected by stray-light;

(d) be non-fluorescent;

(e) show little change in its optical properties with temperature;

(f) not be changed by exposure to normal atmospheres and light; and

(g) be easy to construct and calibrate.

It will be readily appreciated that few materials or methods satisfy all of these requirements. In particular, (g) is often difficult to satisfy and rules out many techniques for all but the national standards laboratories or the most ardent spectroscopist.

5.2 Glass filters

5.2.1 *The NBS glass filters*

A series of four glass filters have been described in a series of papers by NBS staff [1-4]. Typical transmittance curves are shown in

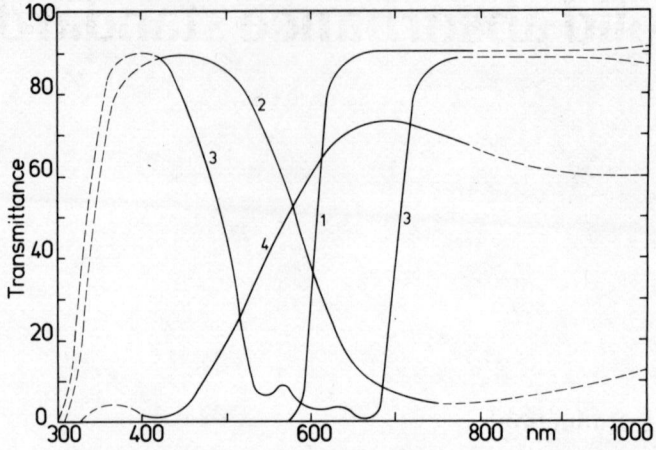

Fig. 5.1   *Transmittance spectra for NBS glass filters: 1: Selenium orange; 2: Copper Green; 3: Cobalt blue; 4: Carbon yellow. From [1].*

Fig. 5.1. Copeland *et al*, [5] studied a Carbon Yellow filter from this series over a period of six years. He concluded that the filter was stable,with a day-to-day variability of ±0.002 absorbance units. The filters were discussed further by Rand [6].

Criticism of these filters centred around the fact that the values supplied with the filters were certified only to ± 3 per cent (0.001 to $0.010A$ depending on the absolute absorbance). The filters all

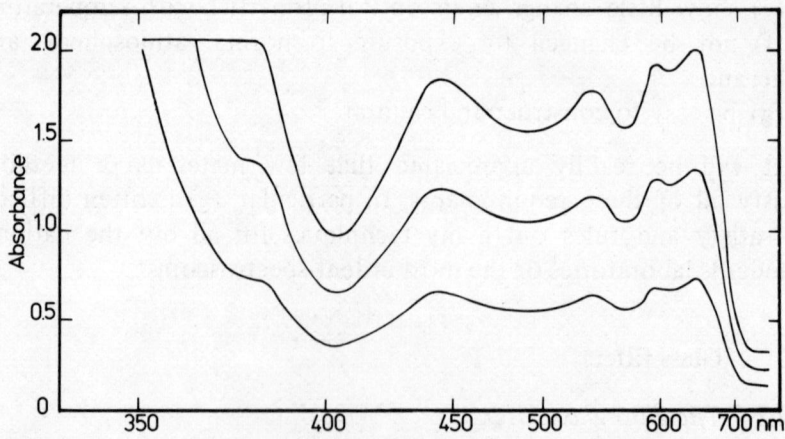

Fig. 5.2   *Absorption spectra of Chance ON 10 filters of various thicknesses, from [7].*

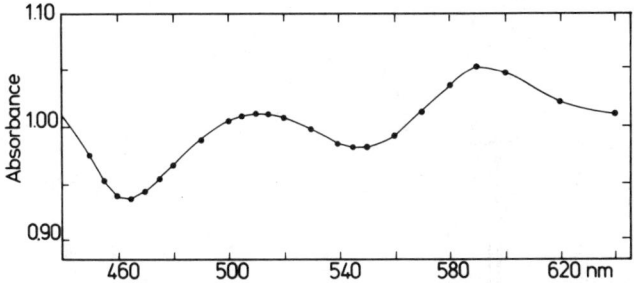

Fig. 5.3   *Absorption spectrum of a Schott NG-4 glass filter, from [11].*

show extensive dependence of absorbance on wavelength setting (at 540 nm the Carbon Yellow curve is rising at a rate of 0.0083 A/nm) and so even minute wavelength setting errors easily resulted in misleading information.

5.2.2   Chance ON10 filters

The use of this material has been discussed by Slavin [7] and also by Porro and Morse [8]. Its absorbance is shown in Fig. 5.2. A particular sample was measured by Slavin 24 times in six different orientations in the beam, and an absorbance of 1.947 with a standard deviation of 0.004 was achieved. To achieve these figures, the material was polished so that the parallelism was within a few micrometres, and the thickness determined to within 1 μm using gauge blocks.

5.2.3   Schott NG-4 glass

Schott NG-4 glass has now become the usual material from which glass absorbance standards are prepared. It is the material used by the NBS in America for SRM 930 and also by the NPL in Britain. The absorbance of NG-4 is shown in Fig. 5.3. The NBS filters are described in a number of reports [9, 10] and comprise a set of three glass filters to give nominal transmittance of 10, 20 and 30 per cent (1.0, 0.7 and 0.5A). Filters are also available from the NPL. These were first described by Sharpe [11] and by Popplewell [12, 13], following a cooperative project between NPL and Pye Unicam, and comprise a set of six filters covering nominal absorbances of 0.1, 0.2, 0.5, 1.0, 1.5 and 1.9A. The filters are manufactured to the most exacting standards and are produced with the filter faces parallel to within 0.5 minutes of arc and the faces flat to within two fringes/cm of mercury green light. They are housed in specially designed stress-free mountings as shown in Fig. 5.4.

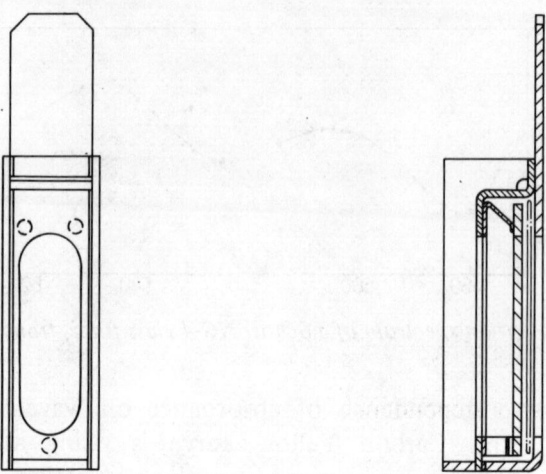

Fig. 5.4   *A stress-free mounting for glass filters, from [11].*

5.2.4   *Stability and ageing effects*

Blevin [14], in 1959, pointed out that the transmittance of Chance ON glass increased over a period of five years. He attributed this to a blooming of the surface and showed that the original transmittance could be restored by gently polishing with cerium oxide.

Both Sharpe [11] and Mavrodineanu and Baldwin [10] investigated the temperature stability of the Schott NG Series. Mavrodineanu concluded that the variation in transmittance with temperature is insignificant at the 0.95 confidence level. For temperature changes of up to $\pm 2°C$, the variation is minimal and averages less than 0.2 per cent of the measured transmittance values.

The NBS also investigated the stability of the glasses when exposed to a high-intensity source and concluded that they possess an acceptable stability. Similar tests on other neutral glasses indicated that these were less stable by a factor of 4 compared with the Schott NG-4 glass.

Initial investigations both by the NBS and by Pye Unicam indicated that the time stability of Schott NG glass was very good. However, in 1976 Drews of Perkin-Elmer noted some instability in NBS filters [15]. It appears that this instability - an anomalous decrease in transmittance of about 1 per cent - has only been noted in one batch of the glass, and the NBS have not found this in subsequent batches.

5.2.5 *Reflection correction*

Back reflections between the filter faces and other elements in a spectrophotometer (e.g. sample compartment windows) can affect the calibration obtained when using a set of neutral filters. A novel method of overcoming this was described by Sharpe [11], who used a clear glass filter to zero the instrument. This filter compensates for the effect of reflections from the surfaces of the test filter, for these are also present when zeroing the instrument.

A rigorous treatment of the numerical correction required is given by Mielenz and Mavrodineanu [16]. However, this correction has to be obtained from measurements on tilted samples in polarized light and is probably beyond the scope of most laboratories.

5.2.6 *Other glasses*

A comprehensive review of some 800 coloured glasses has been published by Dobrowolski *et al.* [17]. This lists a number of neutral glasses which could potentially be used as standards, but little has been published on the various parameters (e.g. stability) which are of paramount importance in selecting a filter for use as a standard.

5.3 The use of polarizers to determine photometric linearity

Polarizers have not found wide favour as standards for a variety of reasons, but some very detailed studies of their use has been carried out by Bennett [18] and by Mielenz and Eckerle [19]. Bennett points out that three polarizers are required if effects due to the polarization of the optical system of the spectrophotometer are to be avoided. Three polarizers are arranged as shown in Fig. 5.5 with the outer two fixed, and the centre one variable. Absorbance ranges of 0-4.0 are readily achievable by rotation of the centre polarizer. With sheet polarizers the device has a useful operating range of

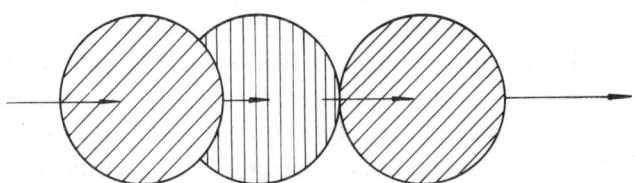

Fig. 5.5 *The principle of the three-polarizer attenuator. The outer filters are fixed with their planes of polarization parallel and the centre one is rotated to vary the transmission.*

375 nm to 700 nm. Linearity can be determined to better than 0.1 per cent. Bennett points out the main factors limiting the accuracy of a three-polarizer transmittance standard to be the extinction ratio of the polarizers, the birefringence of the polarizers, the accuracy with which the outer polarizers can be aligned, and the accuracy with which the middle polarizer can be set. Mielenz and Eckerle take the technique further by using quarter- or half-wave plate attenuators in place of the middle polarizer, thus eliminating the problem of unknown birefringence of the middle plate. However, to achieve accuracies of the order of 0.01 per cent in transmittance, the angular setting of the rotating element must be made to within approximately 0.4 minute of arc. The device is not suitable for routine use.

5.4   Metal screens

Metal screens, in the form either of woven wire screens or of chemically etched thin metal foils, have been investigated by a number of workers [20-24]. Heidt and Bosley [20] reported the use of woven wire screens in 1953 for checking a Cary spectrophotometer, and in 1961 Newman and Binder [21] reported on the use of a series of screens to provide attenuators with an overall range of six orders of magnitude. Interest in the use of attenuators of this type for the calibration of spectrophotometers revived in the early sixties, with the search for a standard that was reasonably neutral throughout the 200-800 nm wavelength range.

Vanderbelt [22] and Slavin [23] studied screens formed by chemically etching thin metal plates. The holes in these screens are conical, of uniform size and spacing, with smooth edges. It is important always to present the screen to the light beam in the same orientation, and Vanderbelt recommended that the smaller diameter end of the pores should always face the light beam. Slavin reported variations of greater than 0.03 absorbance units if the same portion of the screen was used in different instruments of the same design, or if the same instrument was used with different source and detector. He concluded they were not adequate as routine standards, but could be used for expanding the photometric scale in the region near zero transmittance by placing an appropriate screen in the reference beam of the spectrophotometer. Screens are currently used for this purpose and supplied by a number of manufacturers. Typical absorbance curves for a set supplied by Pye Unicam are shown in Fig. 5.6.

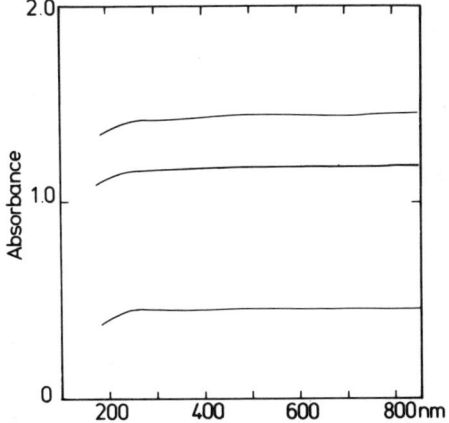

Fig. 5.6   *Three typical absorbance spectra for metal screens.*

5.5   Sector discs

One approach to the problem of accurately varying the amount of radiation falling on the detector is to use an optical chopper to attenuate the beam, rather than an absorbing filter, for example a rotating disc with apertures of known size in it. Hence the length of time that radiation passes through it can be accurately calculated and the 'open-to-shut' time ratio claculated. Such a device was described by Jones and Clarke [25] who pointed out two important considerations. The detector system must average the pulsed flux correctly (i.e. it must obey Talbot's Law) and the method of timing the 'open-to-shut' ratio must be capable of very high accuracy and precision. These devices are not commonly used with modern commercial double-beam spectrophotometers, as the chopping rate could interfere with the normal chopping cycle of the spectrophotometer.

5.6   Metallic filters

One of the major drawbacks of conventional glass neutral density filters is, of course, the fact that they cannot be used in the UV, as all known neutral absorbing glasses have a cut-off edge just below the visible range. Evaporated metal-on-quartz or fused silica has therefore been investigated by a number of workers, notably Mavrodineanu [26] at the NBS and Clarke at the NPL [27]. The major limitation of such filters is the inter-reflections which can

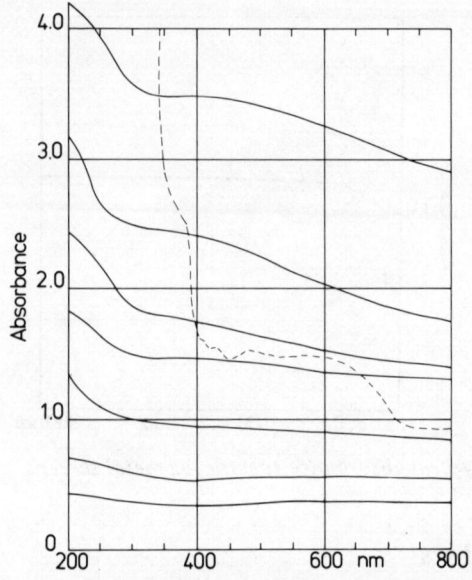

Fig. 5.7   *Typical absorbance spectra for NPL Nichrome-on-fused-silica filters (———) compared with a Schott NG-3 filter (- - - -)* [27]

occur in instruments, resulting from their intrinsic property of attenuating radiation by reflection. However, this apparent drawback can be put to good use, as will be described later.

In his paper, Mavrodineanu describes results obtained with filters of Inconel-on-fused-silica, with the Inconel protected by a clean fused silica plate held in place with an organic cement. Initial results with the NBS high-accuracy spectrophotometer indicated that a positioning error of 3° about the vertical axis could be tolerated. Subsequent evaluation in commercially available spectrophotometers indicated that the filters could be acceptable as transfer standards in spectrophotometry. The NBS has recently announced that they will be selling sets of three chromium-on-quartz filters calibrated over the range 250-635 nm as SRM 2031.

Further work by Clarke and co-workers showed that, by carefully using a particular hard variety of Nichrome alloy, no cover glass is required. Absorbance curves for this material are shown in Fig. 5.7. Also shown is a typical NG-3 filter and it will be seen that there are a number of points where the transmission profiles intersect. Clarke has shown that these points can be used as diagnostic aids as follows.

Let D_m and D_a be the respective instrument errors in absorbance observed for metal film and absorbing glass filters of nearly equal absorbance, and let $R_{(m,a)}$ be the ratio of the metal film reflectance to the reflectance of the absorbing glass filter. The inter-reflective component D_{ra} of the observed error with the absorbing glass filter is then given by:

$$D_{ra} = \frac{D_m - D_a}{R_{(m,a)} - 1}$$

$R_{(m,a)}$ will typically be in the range 5 to 15 for all Nichrome filters of absorbance greater than 0.4. The basic photometric scale error is the simply D_a-D_{ra}. A fuller description of the theory and application is given in [27].

5.7   Light addition methods

In the previous sections, various means of attenuating the beam were considered. All however suffer from some deficiency or other - many, for example, rely on another laboratory having initially calibrated the absorbing material. An alternative approach that is simple - at least in theory - is the light addition methods.

If there are two radiant fluxes X and Y passing through a spectrometer which individually give readings $I(X)$ and $I(Y)$, then when the two fluxes are added, a reading $I(X+Y)$ will be obtained. If the expression

$$\frac{I(X) + I(Y)}{I(X+Y)}$$

is not equal to unity, errors exist and the system is non-linear.

The technique was probably first used by Elster and Geilet [28] in 1893 and many workers have since produced variations on the theme [30-32]. The most important of these variations are: (i) supplementary light methods as used by Reule [29], and (ii) aperture methods using various apertures that are opened and closed separately or in combination Clarke [30], Douglass and Emary [31], Mielenz and Eckerle [32].

Of these, the second approach is probably preferable in that a second light source is not required and the apparatus can be relatively simple - a schematic diagram of such a device is shown in Fig. 5.8. These devices form the basis of the checking carried out by both the

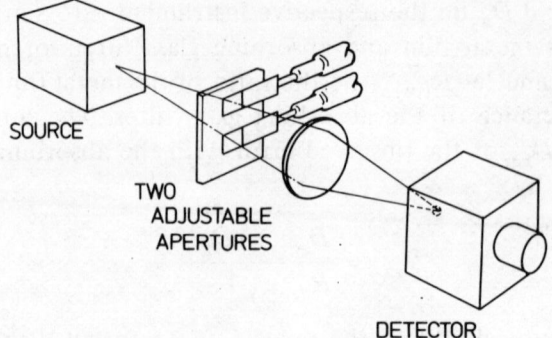

Fig. 5.8   *Schematic diagram of the double-aperture method.*

NPL and the NBS on their high-accuracy spectrophotometers before calibrating and issuing calibrated Schott filters. A device working on a very similar principle had been developed some years previously by Gilford Instruments [31]. This involved two light-integration areas, coupled by ten apertures, so giving a large range of calibration points. In general these techniques are probably beyond the capabilities and needs of most users, and the excellent papers produced both by the NPL and the NBS should be consulted for further details [30, 32].

5.8   Conclusions and recommendations

For routine use, the recommendation must be a set of neutral Schott glasses, as supplied by NPL or NBS, or by a number of the major instrument manufacturers. These should be submitted for recalibration at yearly intervals.

Alternatively, evaporated metal-on-quartz filters may be used. These have the advantage that they can be used to check absorbance accuracy in the UV as well as the visible region. Their main advantage, however, lies in their use in tracking down the effects of back reflections within the instrument.

Finally, comment should be offered on the relative merits of solutions versus solid filters. The author, must agree with the comments expressed by Clarke [27]: 'The liquid standard tests the complete organization in a chemistry laboratory, if the solution is to be made up to a specified concentration and purity. On the other hand, solid spectrophotometric standards, if stable, can be used to test specifically the intrinsic performance of the instrument, if conditions of use are tightly controlled. . . .' 'If one wants to consider the actual photomeasurement capability of the instrument apart from considerations of support-staff competence, liquid and cuvette handling procedures or the quality of the cuvettes involved, then solid transmittance standards are the indicated choice'.

References

1 Gibson, K.S. (1949), NBS Circular 484.
2 Gibson, K.S., Walker, G.K. and Brown, M.E. (1934), *J. Opt. Soc. Amer.,* **24**, 58.
3 NBS Circular LC1017 10 (1955).
4 Gibson, K.S. and Balcom, M.M. (1947), *J. Res. Nat. Bur. Std.,* **38**, 601; also (1947) *J. Opt. Soc. Amer.,* **37**, 593.
5 Copeland, B.E., King, J. and Willis, C. (1968), *Amer. J. Clin. Path.,* **49**, 459.
6 Rand, R.N. (1969), *Clin. Chem.,* **15**, 839.
7 Slavin, W. (1962), *J. Opt. Soc. Amer.,* **52**, 1399.
8 Porro, T.J. and Morse, H.T. (1966), Pittsburg Conference on Analytical Chemistry.
9 Mavrodineanu, R. (1971), NBS Technical Note 584.
10 Mavrodineanu, R. and Baldwin, J.R. (1975), NBS Special Publication 260-51.
11 Sharpe, M.R. (1975), *UV Spec. Grp. Bull.,* **3**, 57.
12 Popplewell, B.P. (1975), *Measurement focus* (British Calibration Service Newsletter, **6**, 1.
13 Popplewell, B.P. (1977), *UV Spec. Grp. Bull,* **5**, 90.
14 Blevin, W.R. (1959), *Opt. Acta,* **6**, 99.
15 Mavrodineanu, R. and Drews, U.W. (1976), *Clin, Chem.,* **22**, 1230.
16 Mielenz, K.D. and Mavrodineanu, R. (1973), *J. Res. Nat. Bur. Std.,* **77A**, 699.
17 Dobrowolski, J.A., Marsh, G.E., Charbonneau, D.G., Eng. J. and Josephy, P.D. (1977), *Appl. Opt.,* **11**, 594.
18 Bennett, H.E. (1966), *Appl. Opt.,* **5**, 1265.
19 Mielenz, K.D. and Eckerle, K.L. (1972), *Appl. Opt.,* **11**, 594.
20 Heidt, L.J. and Bosley, D.E. (1953), *J. Opt. Soc. Amer.,* **43**, 760.
21 Newman, P.A. and Binder, R. (1961), *Rev. Sci. Instr.,* **32**, 351.
22 Vanderbelt, J.M. (1962), *J. Opt. Soc. Amer.,* **52**, 284.
23 Slavin, W. (1962), *J. Opt. Soc. Amer.,* **52**, 1399.
24 Bryan, F.R. (1963), *Appl. Spectr.,* **17**, 19.
25 Jones, O.C. and Clarke, F.J.J. (1951), *Nature,* **191**, 1290.
26 Mavrodineanu, R. (1976), *J. Res. Nat. Bur. Std.,* **80A**, 637.
27 Clarke, F.J.J. (1977), *UV Spec. Grp. Bull.,* **5**, 104.
28 Elster, J. and Geilet, H. (1893), *Wied. Ann.,* **48**, 625.
29 Reule, A. (1968), *Appl. Opt.,* **7**, 1023.
30 Clarke, F.J.J. (1972), *J. Res. Nat. Bur. Std.,* **76A**, 375.
31 Douglass, S. and Emary, R. (1976), *Lab. Equip. Digest.,* December, 57.
32 Mielenz, K.D. and Eckerle, K.L. (1972), *Appl. Opt.,* **11**, 2294.

6 Stray-light

6.1   Introduction

It has been rightly said that 'of all the spectrophotometric parameters, the one that probably causes the most confusion is stray-light' [1]. Stray-light is a likely source of error in the measurement of absorbance, and many of the apparent deviations from the Beer-Lambert Law are due to its presence [1-10]. Until cheap tunable laser sources become available, stray-light will continue to be a problem, particularly when high absorbances are being measured. The precise measurement of stray-light present in a spectrophotometer and of the errors it introduces are tedious processes, unlikely to be carried out in the majority of laboratories. Fortunately, stray-light is easily detected, at least in the spectral regions where it matters most, and simple considerations will usually suffice to show whether or not it is likely to affect absorbance measurements on a given sample. It must be emphasized that the effect of stray-light is a function of many factors, not least the sample itself. Furthermore, the amount of stray-light in a given spectrophotometer increases slowly with time, due to deterioration of the source and optics.

In this chapter, light entering through light-leaks is excluded from the term 'stray-light'. Light-leaks should, of course, be absent in any properly constructed spectrophotometer. Their presence can be detected by covering the spectrophotometer, and noting any change in the absorbance reading or reduction of the noise level. Another factor which will be excluded is fluorescence of the sample. This can be detected by a change in absorbance reading occurring when the sample is placed at a different distance from the detector. Although fluorescence is quite common, it is usually very weak and is unlikely to have an effect except at high absorbance readings.

6.2 Definitions

Stray-light, in general terms, is radiation of wavelengths outside the narrow band norminally transmitted by the monochromator (the 'passband') of the monochromator). For the purpose of this report two terms will be used:

(a) Monochromator stray-light (MSL)
(b) Instrumental stray-light (ISL)

All monochromators transmit a small amount of radiation outside their passband; the ratio of this small amount to the total amount of radiation transmitted by the monochromator will be referred to as 'monochromator stray-light'. The amounts of stray-light and the light within the passband are modified during their passage through the rest of the optical system (including the sample). Furthermore, the electrical signal produced by the detector is not directly related to the amounts of stray-and wanted-light incident on the detector, because the detector's sensitivity is a function of wavelength. For these reasons the term 'instrumental stray-light', introduced by Slavin [1] is a useful one. It can be defined as 'the ratio of the signal produced by the detector for radiation of all wavelengths outside the monochromator passband, to the total signal at a particular wavelength setting'. It is this ratio which determines the effect of stray-light on the absorbance reading.

6.3 Origin and effects of stray-light

6.3.1 *Monochromator stray-light (MSL)*

A perfect monochromator would transmit light only within its passband; light of other wavelengths entering the monochromator would be totally absorbed. In practice, scattering and diffraction inside the monochromator introduce light of other wavelengths, i.e. stray-light, into the exit beam. Imperfections in the optical surfaces are largely responsible and, as far as a well-designed diffraction grating monochromator is concerned, nearly all the stray-light originates at the grating [11-14]. Sharpe and Irish [14] in a theoretical and experimental study have shown that the stray-light is due mainly to random errors in the spacing and in the depth of the grooves. Gratings produced by a holographic process (such gratings are now available in many commercial spectrophotometers) produce less stray-light, by about a factor of ten, than gratings produced mechanically. It should be noted, however, that even a perfect grating would produce some stray-light [14].

Stray-light of wavelengths close to the wavelength setting of the monochromator is referred to as 'near' stray-light, and the remainder of the stray-light is called 'far' stray-light. Prism monochromators may give a lower level of far stray-light compared with grating monochromators, but their poorer dispersion produces a high level of near stray-light. The overall stray-light performance of modern grating monochromators is greatly superior to that of prism monochromators [15]. Further details of the origins of MSL are given in references [15] and [16]. The latter reference also deals with the problems of specifying the stray-light performance of grating monochromators.

Obviously, MSL also depends on the variation of intensity with wavelength of the source. Thus when working in the region 320-400 nm, a deuterium lamp gives rise to less MSL than a tungsten lamp because the intensity of the latter is much greater at wavelengths greater than 400 nm. The monochromator parameters: wavelength setting, slitwidth, and slitheight are further factors which influence MSL. In the case of a single monochromator, both the intensities of the stray-light and the light within the passband are proportional to the square of the slitwidth, so MSL is independent of the width. It is approximately proportional to slitheight [10], though in most spectrophotometers no means of varying the height is provided. In the case of a double monochromator, MSL is very much reduced, but it is proportional to slitwidth and to the square of the slitheight [10]. The effect of wavelength setting is described in Section 6.4.1.

6.3.2 *Instrumental stray-light (ISL)*

As already emphasized, the monochromator stray-light is modified by other components of the spectrophotometer, and by the sample itself. The following factors are involved:

(a) The condition of the mirrors (outside the monochromator) and any other optical components. The reflectance of an aluminized mirror is high in the visible and UV regions, but is liable to deteriorate with time, particularly below 250 nm. In the latter region much of the stray-light is of longer wavelengths, therefore its proportion is likely to increase with time.

(b) The absorption spectrum of the sample (including the solvent and the cell). The amounts of stray- and wanted-light are invariably affected in different ways by the sample. The proportion of stray-light can be increased or decreased (usually increased).

(c) Below 200 nm absorption by atmospheric oxygen becomes important. In this region nearly all the stray-light is of longer wavelengths, hence its proportion is increased.

Finally, the instrumental stray-light will be determined by the sensitivity-versus-wavelength curve of the detector. Typical photo-multipliers used in spectrophotometers have a fairly constant sensitivity in the range 190-400 nm, but a steady decrease in sensitivity at longer wavelengths. The low sensitivity just beyond the red end of the visible spectrum ($\lambda > 700$ nm) is most serious as far as ISL is concerned. At 850 nm, for example, the quantum efficiency of an extended S-20 photo-cathode is 1 per cent, compared with 20 per cent at 400 nm.

6.3.3 *Variation of ISL with wavelength*

Typical results of all the factors described in Sections 6.3.1 and 6.3.2 are as follows:

(a) *Tungsten lamp range (usually 320-850 nm)*
The variation of ISL with wavelength is dominated by two factors: the variation with wavelength of the tungsten lamp output, and the low sensitivity of most photomultipliers above 700 nm. The peak output of the lamp is at ca. 1000 nm and the output falls rapidly at shorter wavelengths (the well known black-body radiation curve is followed approximately). Potentially, ISL can be appreciable at wavelengths below 400 nm and above 800 nm, but in practice it is reduced considerably by stray-light filters (see Section 6.3.5).

(b) *Deuterium lamp range (usually 190-320 nm).*
Several factors are important in this range:

(i) MSL increases appreciably below ca. 250 nm (the fall in output of the source at shorter wavelengths as it ages is an important contributory factor);

(ii) The condition of the mirrors;

(iii) Absorption by the sample, particularly when a solvent is used at wavelengths below its cut-off;

(iv) Below 200 nm absorption by atmospheric oxygen becomes important.

The general effect of (i)-(iv) is to increase ISL as wavelength is decreased from about 250 nm. It is in this region that the effect of ISL is likely to be most serious.

6.3.4   *Errors introduced by stray-light*

It is important to remember that ISL is a function of the sample. A general statement of the errors introduced cannot be made, even if the ISL in the absence of the sample is known. The equations derived below will emphasize this point:

For a given wavelength setting, λ, let:

P_t = total photomultiplier signal, in absence of cell + sample.

P_λ = contribution to P_t made by light within the monochromator passband.

P_s = contribution to P_t made by stray-light.

P'_t = total photomultiplier signal with cell + sample.

T_λ = true transmittance of cell + sample at wavelength λ

T'_λ = apparent transmittance of cell + sample at wavelength λ

T_s = transmittance of cell + sample for the stray-light to which the detector is sensitive.

s = instrumental stray-light, ISL, in the absence of a sample.

In the absence of the cell + sample:

$$P_t = P_\lambda + P_s$$

In the presence of the cell + sample, assuming a linear response of the photomultiplier:

$$P'_t = P_\lambda T_\lambda + P_s T_s$$

Apparent transmittance $T'_\lambda = \dfrac{P_\lambda T_\lambda + P_s T_s}{P_t}$

But $P_s = sP_t$ and $P_\lambda = (1\text{-}s)P_t$:

$$T'_\lambda = (1\text{-}s)T_\lambda + sT_s$$
$$= T_\lambda + s(T_s - T_\lambda) \tag{6.1}$$

It is clear from Equation (6.1) that the apparent transmittance can be greater than or less than the true transmittance, depending on whether T_s, the transmittance of stray-light by the sample, is greater than or less than T_λ, the true transmittance. Note that 'stray-light'

Table 6.1: *The effect of ISL upon the apparent absorbance of a sample, assuming that no stray-light is absorbed by the sample.*

True absorbance A_λ	Percentage of ISL		
	0.01	0.10	1.00
0.500	0.500	0.499	0.491
1.000	1.000	0.996	0.963
1.500	1.499	1.487	1.384
2.000	1.996	1.959	1.701
2.500	2.486	2.381	1.882
3.000	2.959	2.699	1.959

in this context includes only the stray-light to which the photomultiplier is sensitive.

If we consider the region where stray-light effects are most likely to be appreciable, i.e. $\lambda < 250$ nm, the usual situation is that the sample absorbs strongly in this region, but its absorbance is low at longer wavelengths. Most of the stray-light is also of longer wavelengths, hence $T_s > T_\lambda$ and $T_\lambda' > T_\lambda$, i.e the apparent absorbance is *low*. This is the commonest effect of stray-light. If we assume that *all* the stray-light is transmitted by the sample, i.e. $T_s = 1$, then:

$$T_\lambda' = T_\lambda (1 - s) + s \qquad (6.2)$$

$$\text{i.e. } A_\lambda' = -\log_{10} [10^{-A}\lambda (1 - s) + s] \qquad (6.3)$$

where A_λ' and A_λ are the apparent and true absorbances of the sample. Equation (6.3), or an equivalent one, has been quoted several times [1-5], and graphs of the absorbance error as a function of absorbance for various values of s drawn [1-4], 7, 8]. Table 6.1 shows some values of A_λ' for 0.01, 0.1, and 1.0 per cent ISL.

Note that 'sample' in this context means cell + solvent + solute. The figures can be applied to the solute alone if the absorbances of the cell and solvent are negligible.

Plots of apparent absorbance against true absorbance according to Equation (6.3) have the form shown in Fig. 6.1. The apparent absorbance reaches a limiting value given by $-\log_{10} s$.

The situation in which $T_s < T_\lambda$, giving a *high* apparent absorbance, is unusual, but can occur when an absorbance minimum of a sample is being measured.

When a double-beam spectrophotometer is used, and a com-

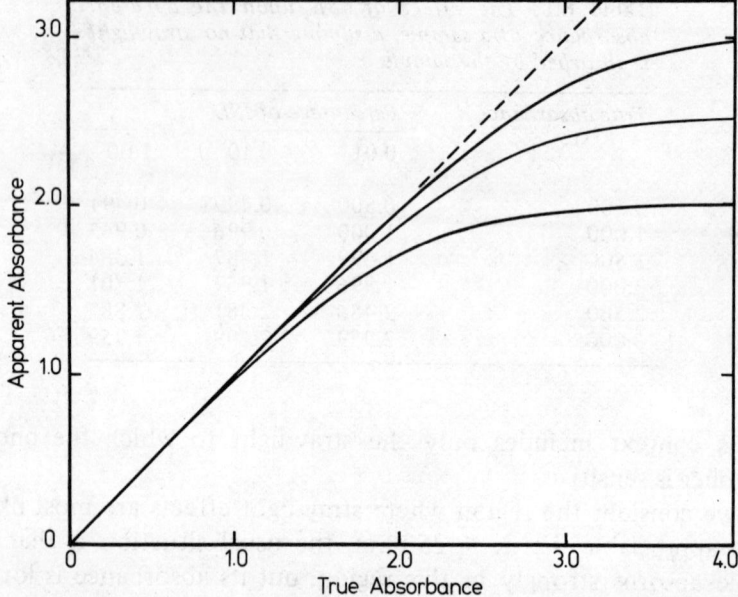

Fig. 6.1   *Plots of apparent absorbance (A') against true absorbance (A) for different percentages of stray-light.*

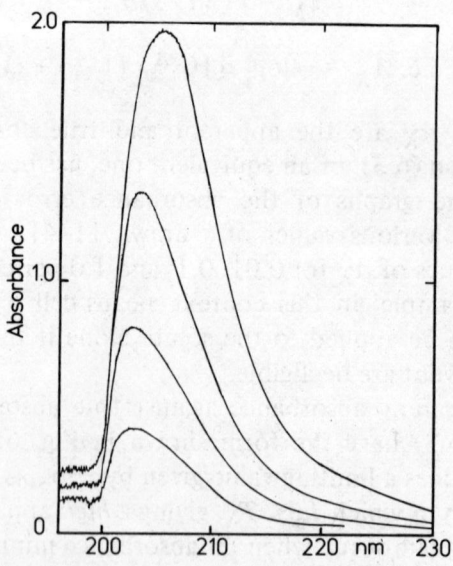

Fig. 6.2   *Absorption spectrum of maleic acid in ethanol at various concentrations, showing the effect of stray-light.*

pensating cell containing solvent is placed in the reference beam, it is tempting to think that all effects due to the solvent are cancelled out. This is emphatically not the case as far as stray-light is concerned. For example, consider chloroform which cuts-off at about 250 nm. If an attempt is made to measure the absorbance of a solute at, say, 240 nm, the solvent increases the proportion of stray-light considerably (the stray-light being of longer wavelengths, is therefore freely transmitted). In extreme cases stray-light will dominate the result. Similar effects can occur below 200 nm, due to absorption by atmospheric oxygen.

It is clear that stray-light can cause apparent deviations from Beer's Law. In the usual situation, $T_s > T_\lambda$, the deviations are negative, and plots of apparent absorbance against concentrations are similar to Fig. 6.1.

When a spectrum is scanned, stray-light can produce a characteristic distortion of band shape if its amount is changing over the region of the band. An associated effect is displacement of the band maximum. Figs 6.2-6.4 illustrate these points. Fig. 6.2 shows spectra of maleic acid in ethanol at four different concentrations. The ISL in the spectrophotometer used for the spectra was *ca.* 1 per cent at 200 nm in the absence of a sample, but absorption by the ethanol below 215 nm brought about a large increase in ISL (see Fig. 6.3 for the

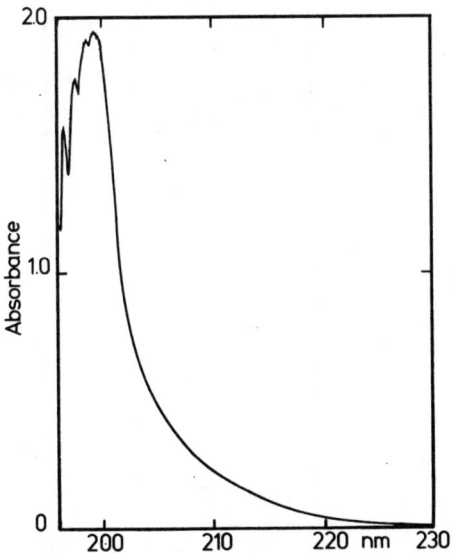

Fig. 6.3   *Absorption spectrum of ethanol showing the effect of stray-light.*

spectrum of the solvent). The depression of the absorbance below 215 nm produced the characteristic 'cut-off' appearance of the bands. Associated effects are the shift of the band maximum, and the high noise level below 200 nm. The spectrum of the solvent also shows stray-light effects; below 200 nm the trace should be off-scale, but the depression of the absorbance has caused a false peak to appear. The sharp downward pointing 'peaks' are due to atmospheric oxygen; ISL is increased at wavelengths of strong oxygen absorption, as noted above. Fig. 6.4 shows spectra of maleic acid in water. The absorbance of water is much lower than that of ethanol in this region, consequently the ISL is only increased slightly. The bands have a normal shape, and the position of the maximum does not change with concentration.

If ISL is appreciable but constant over the range of a band, then its effect is to depress the top of the band. A diagram showing computer simulations of this effect is shown in reference [8].

Broad bands, such as those usually obtained with liquid samples, are affected little, if at all, by near stray-light, for this is absorbed by the sample. On the other hand, narrow bands are liable to be affected considerably by both near and far stray-light.

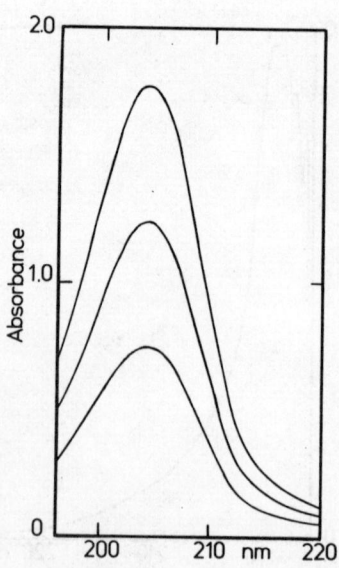

Fig 6.4   *Absorption spectra of maleic acid in water at various concentrations.*

6.3.5　*Reduction of stray-light*

(a) Most spectrometers are equipped with a stray-light filter (e.g. Corning 9863, Schott UG5 or Hoya U-330; Chance OX7 is no longer made) which is placed in the light beam at wavelengths below 400 nm in the range of the tungsten lamp. Such a filter absorbs most of the visible light but transmits well in the near-UV. When the spectrophotometer is set to 350 nm, for example, most of the stray-light is of wavelengths in the visible region, hence the stray-light level is reduced by the filter. Some spectrophotometers also use the Chance OX7 as a stray-light filter above 700 nm, where ISL increases due to the fall-off in the photo-multiplier sensitivity. A more effective filter for the region $\lambda > 750$ nm is the Wratten 88A which is practically opaque to visible light.

A stray-light filter which transmitted freely below *ca.* 220 nm would be very useful, but hitherto no filter suitable in practice has been devised. A liquid film of sodium is transparent below 210 nm [17].

(b) Under circumstances where ISL is appreciable and cannot be reduced, the error can be minimized by working at low concentrations. Measurements can be made at several concentrations, and molar absorptivities plotted against concentration. Extrapolation to zero concentration will yield a corrected value of the molar absorptivity [18]. If solvent cut-off is a factor, a reduction in the cell path-length will extend the range over which stray-light errors can be neglected; this is particularly useful in the region below 250 nm where molar absorptivity are generally high and path-lengths of 1 mm or less are practicable - see for example [19]. Sometimes it is possible to use a different solvent having a shorter cut-off wavelength (note that traces of impurities, or sometimes the presence of oxygen, can affect solvent cut-off markedly).

(c) If measurements are made below 200 nm, the stray-light level can be reduced by flushing as much as possible of the optical path of the spectrophotometer with nitrogen. Some of the problems encountered when working in this region have been discussed by Turner [20].

6.4　**Measurement of stray-light**

6.4.1　*Measurement of monochromator stray-light (MSL)*

The general method for the measurement of MSL is to illuminate the entrance slit of the monochromator with a monochromatic source, and determine the intensity of the emergent beam as a function of wavelength setting. Most of the light emerges at wavelength settings within the monochromator passband; light emerging at other settings is stray-light. Effectively, the transmission curve of the monochromator is determined. A complete study of MSL requires measurements with several monochromatic sources covering the wavelength range of the monochromator. To obtain the MSL as defined in Section 6.2.2, the results have to be combined with the variation of output of the normal source (tungsten or deuterium lamp) with wavelength, and an integration carried out over a sufficiently large wavelength range.

The measurement of MSL is a tedious process requiring specialized equipment, and only a few such measurements have been reported. Two methods have been used to obtain the monochromatic incident beam. In the first of these the source is either a continuous one [21] or an arc emitting several lines (e.g. mercury) [22, 23]. Light from the source is passed through a double monochromator, and the nearly monochromatic emergent light passed into the monochromator under test. In the second method the light source is a laser, and a

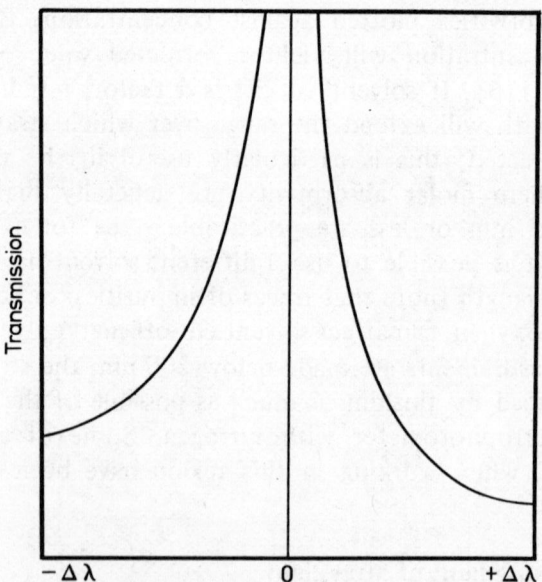

Fig. 6.5   *A typical monochromator transmission curve. Minor peaks and irregularites are not shown. The abscissa, $\Delta\lambda$, is the difference between the wavelength setting of the monochromator and the wavelength of the monochromatic light passing through it.*

double monochromator is not required [14, 16, 24].

A typical monochromator transmission curve [23] is sketched in Fig. 6.5. Away from the monochromator passband the profile is approximately exponential (see Reference [24] for curves plotted on a logarithmic scale). Most of the monochromators investigated by Tarrant [23] gave transmission curves whose shape is practically independent of the wavelength of the monochromatic light. This is in agreement with the work of Rossler [24] who introduced a 'figure of merit', $\log I/U$, where I = intensity of transmitted light at the maximum of the transmission curve, and U = the intensity at the beginning of the exponential sections. Rossler found that $\log I/U$ for a particular monochromator was almost independent of the wavelength of the monochromatic light, and of the slitwidth of the monochromator. Tarrant showed that the height of the transmission curve increases with decreasing wavelength, and that a height factor of the form $h = A + B/\lambda^2$, where A and B are constants for a given monochromator, gives a reasonably good correlation. By normalizing a set of transmission curves, Tarrant obtained a mean shape which he called the 'figurine' of the monochromator. Then the stray-light intensity relative to the maximum is given as a function of wavelength (λ) of the monochromatic light, and wavelength setting (x) by:

$$M(\lambda, x) = (A + B/\lambda^2) f(x)$$

where f(x) is the figurine, a function of x only.

6.4.2　*The use of cut-off filters*

This involves measuring the absorbance of a filter just beyond its cut-off point [1-3, 6-8, 10, 25-27, 28]. The filter must have an absorbance curve which rises sharply over a narrow wavelength range. Fig. 6.6 illustrates the principle of the method. The solid curve represents the true absorbance of the filter, and the dashed curve the apparent absorbance in the presence of stray-light (a typical situation is shown, i.e. ISL increasing as wavelength decreasing). If A' is the apparent absorbance at a wavelength λ' where the true absorbance is very high, then the corresponding transmittance T' is the ISL at λ'. The ISL so determined is a function of the filter's absorbance curve, and can only be taken as a rough indication of the ISL present when a different absorber is in the beam. However, as a means of detecting stray-light the method is usually satisfactory, and it is the one recommended for routine use (see Section 10.3.1 for specific recommendations).

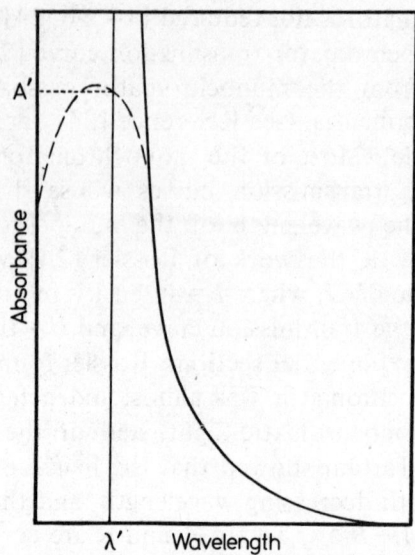

Fig. 6.6   The use of a cut-off filter to measure stray-light. The continuous curve is the absorbance of the filter, and the dashed line is the apparent absorbance curve.

With reference to Fig. 6.6, stray-light of wavelengths less than λ' is absorbed by the filter, and a certain amount of the 'near' stray-light of wavelengths greater than λ' is also absorbed. If the absorbance is measured at a wavelength appreciably below λ', then a greater amount of stray-light is absorbed, and an optimistic impression of the level of stray-light may result. Therefore, the wavelength at which stray-light is being measured should always be close to the filter cut-off.

Although the value of the ISL obtained by using a filter cannot be equated to the ISL present when a different absorber is in the beam, it is safe to assume that if 0.01 per cent or less ISL is detected by a filter, then measurements on any sample will be unaffected, at least up to 1.0 absorbance units, provided that the absorbance of the cell and solvent is negligible (see Section 6.3.4). Equation (6.3) can be used to give an approximate idea of the absorbance error. This equation assumes that no stray-light is absorbed, whereas both cut-off filters and samples absorb some stray-light. In most cases of solution spectra the bands are broad, and samples absorb more stray-light than filters. Therefore Equation (6.3) will usually overestimate the error.

As far as the deuterium lamp region is concerned (usually $\lambda < 320$ nm), ISL is normally measured at the low end of the wavelength range where its amount is greatest. Therefore a low ISL reading obtained with a cut-off filter at, say, 210 nm indicates a yet lower amount at longer wavelengths. In the tungsten lamp region, a low ISL reading at *ca.* 340 nm also indicates a lower amount at longer wavelengths, at least up to 700 nm. A separate test for ISL should be carried out at the long-wavelength end, because the low photo-multiplier sensitivity increases ISL. As already noted, most spectrophotometers incorporate stray-light filters in the ranges 320-400 nm and >750 nm (in the absence of such filters, ISL can be considerable in these ranges).

6.4.3 *Deviations from the Beer-Lamber Law*

Measurements of absorbance are carried out on a series of solutions of different concentrations of a given solute, the pathlength being kept constant [14]. Alternatively, the pathlength can be varied and the same solution used for all the measurements [29]. In the former case one must be able to rule out other causes of deviations from the Beer-Lambert Law, e.g. association of the solute, effect of finite slitwidth on a narrow band.

The deviations due to stray-light are usually negative but can be positive if sufficient stray-light is absorbed by the sample (Section 6.3.4). Equation (6.3) may be found to apply, but any value of ISL so found is of little significance because the equation assumes no absorption of stray-light by the sample. Clearly the method is of little value from a quantitative viewpoint; it is best regarded as giving an indication of the presence of stray-light, and providing a calibration curve for the particular solute (and solvent).

Miranda and Conte [30] proposed a more complex method based on deviations from the Beer-Lambert Law. They suggest measuring the absorbances of a series of solutions whose concentrations are in an arithmetic series. They claim that their method is better than the cut-off filter method when stray-light is high, but the limitations noted in the previous paragraph still apply, and the complexity of their method does not commend it.

6.4.4 *Other methods*

(a) 'Total absorber' method. If a sample has a very strong, narrow absorption band, then the apparent minimum transmittance will be a close approximation to the ISL in the absence of a sample, provided that no other absorption bands are present. The width of the band must be close to the width of the monochromator passband. This method is of very limited application in the UV and visible regions on account of the lack of suitable absorbing materials [3]. Tunnicliff [31] used mercury vapour which is an almost ideal total absorber at 253.7 nm. No other absorption bands occur above 210 nm, and the width of the band can be varied by changing the temperature. Less ideal total absorbers are glasses or solutions containing rare-earth ions [3].

(b) Measurement of absorbance with and without a band-pass filter [2, 5]. The filter blocks the stray-light, and any absorbance change is a measure of ISL. The result is a function of the absorbing material, and the absorption curve of the filter.

(c) Variation of slitheight. As pointed out in Section 6.3.1, MSL is approximately proportional to slitheight in a single monochromator. The proportionality holds more accurately for a prism monochromator; with gratings the stray-light is predominantly in directions perpendicular to the grooves, so MSL varies according to a fractional power of slitheight [10]. Preston [32] devised a method for the measurement of MSL which involves fitting a plate over the entrance slit to cover a little more than half its height. A similar plate is arranged to cover the half of the exit slit which is illuminated. Under these conditions the wanted light is blanked, but half of the stray-light emerges. Reule [27] has discussed this method critically, and points out that any method based on the variation of slitheight requires one to assume that all parts of the slit contribute uniformly to the radiant flux. This requirement is hardly ever met, so reduction of slitheight is not a sound basis for the measurement of stray-light, though it is sometimes useful for reducing it.

6.5   Theoretical treatment of stray-light

As pointed out in Section 6.3.1, in a well-designed diffraction grating monochromator nearly all the stray-light originates at the grating. The principal grating imperfections which cause stray-light are [33]:

(a) Random errors in the spacing of the grooves;
(b) Random errors in the depth of the grooves;
(c) Small-scale random surface roughness.

Sharpe and Irish [14] have treated the effects of these imperfections theoretically, and have derived an equation for the total MSL in terms of parameters which describe the magnitudes of the imperfections.

The equation was fitted to measurements of MSL made using a monochromatic source, and reasonable values of the parameters were obtained. The relative contributions of the imperfections (a)-(c) were shown to vary considerably with the wavelength of the monochromatic light and the wavelength setting of the monochromator. Sharpe and Irish also derived an equation for the apparent transmittance of a filter when it is placed in a spectro-photometer having a continuous source. Good agreement was obtained between the predicted and observed transmittances of a series of Schott glass cut-off filters.

References

1 Slavin, W. (1963), *Anal. Chem.,* **35**, 561.
2 Holiday, E.R. and Beaven, G.M. (1950), *Photoelec. Spec. Group Bull.,* **3**, 53.
3 Goldring, L.S., Hawes, R.C., Hare, G.M., Beckman, A.O., and Stickney, M.E. (1953), *Anal. Chem.,* **25**, 869.
4 Lothian, G.F. (1963), *Analyst,* **88**, 678.
5 Menzies, A.C. (1960), *Pure Appl. Chem.,* **1**, 147.
6 Hartree, E.F. (1963), *Photoelec. Spec. Group Bull.,* **15**, 398.
7 Rand, R.N. (1969), *Clin. Chem.,* **15**, 839.
8 Cook, R.B. and Jankow, R. (1972), *J. Chem. Educ.,* **49**, 405.
9 Beeler, M.F. and Lancaster, R.G. (1975), *Amer. J. Clin. Path.,* **63**, 953.
10 ASTM standard E-387 (Annual book of ASTM standards, 1978).
11 Welford, W.T. (1965), *Prog. Optics,* **4**, 241.
12 Pribham, J.K. and Penchina, J.K. (1968), *Appl. Optics,* **7**, 2005.
13 Leinert, C. and Kluppelberg, D. (1974), *Appl. Optics,* **13**, 556.
14 Sharpe, M.R. and Irish, D. (1978), *Optica Acta,* **25**, 861.
15 Brown, S. and Tarrant, A.W.S. (1978), *Optica Acta,* **25**, 1175.
16 Verrill, J.F., *Optica Acta* (to be published).
17 Shaw, C.H. and Foreman, W.T. (1959), *J. Opt. Soc. Amer.,* **49**, 724.
18 Fog, J. (1962), *Nature,* **193**, 564.
19 Buist, G.J. and Tabatabai, S.M. (1979), *J. Chem. Soc. Far. Trans. 1,* **75**, 631.
20 Turner, D.W. (1962), *Photoelec. Spec. Group Bull.,* **14**, 388.
21 Donaldson, R. (1950), *Photoelec. Spec. Group Bull.,* **3**, 45.
22 Pritchard, B.S. (1955), *J. Opt. Soc. Amer.,* **45**, 356.
23 Tarrant, A.W.S. (1978), *Optica Acta,* **25**, 1167.
24 Rossler, F. (1974), *Optik,* **41**, 293.
25 Poulson, R.E. (1964), *Appl. Optics,* **3**, 99.
26 West, M.A. and Kemp, D.R. (1976), *Int. Laboratory,* May/June, 27.
27 Reule, A.G. (1976), *J. Res. Nat. Bur. Std.,* **80A**, 609.
28 Francis, R.J. (1979), *Stray-light, silica coatings, and master holographic gratings,* Pye-Unicam Ltd, Cambridge, England.

29 Hogness, T.R., Zscheile, F.P. and Sidwell, A.E. (1937), *J. Phys. Chem.*, **41**, 379.
30 Miranda, C. and Conte, P. (1971), *Appl. Spectr.* 1971, **25**, 557.
31 Tunnicliff, D.D. (1955), *J. Opt. Soc. Amer.*, **45**, 963.
32 Preston, J.S. (1936), *J. Sci. Instr.*, **13**, 368.
33 Palmer, E.W., Hutley, M.C., Franks, A., Verrill, J.F. and Gale, B. (1975), *Rep. Prog. Phys.*, **38**, 975.

7 Wavelength calibration

7.1 Introduction

All spectrometers should be checked regularly for wavelength accuracy for, despite the claims of manufacturers, prism and grating mountings and the driving mechanisms are susceptible to dirt, vibration and the effects of thermal expansion.

Many standards have been used in the past, each having its own particular merits, but few have withstood the test of time. The successful methods are still in use because of their simplicity and general reliability. A good standard should:

  (a) be readily available;
  (b) be easy to assemble or prepare;
  (c) be easy and safe to handle;
  (d) suit the optical properties of the instrument, e.g. spectral bandwidth, etc.;
  (e) be unaffected by environmental conditions, e.g. have a small temperature coefficient, have good chemical stability, etc.

  The recommended methods fall into two categories: (i) the measurement of atomic emission lines from vapour discharge lamps, and (ii) the location of maxima in the absorption spectra of glass filters or solutions with very narrow absorption bands.

7.2 Line source standards

The most general and accurate calibration method is to introduce a discharge lamp into the lamp housing of the spectrometer. Table 7.1 lists some elements that have useful emission lines in the range 185 to 800 nm. Precise values for the location of these lines can be obtained from reference works, but it should be borne in mind

Table 7.1: *The useful spectral ranges of selected vapour discharge lamps for wavelength calibration. Precise values can be obtained from [17].*

Element	Useful wavelength range (nm)
Mercury Zinc Cadmium	185-400
Magnesium Zinc Lead Cadmium Copper Mercury Rhodium	203-420
Neon Tungsten Calcium Strontium Lithium Potassium Sodium	400-800
Krypton	up to 892.9
Argon	up to 811.5

Table 7.2: *The positions of principal neon emission lines* in vacuo *taken from Menzies [1].*

Wavelength (nm)
533.1
534.1
540.1
585.3
594.5
614.3
633.4
640.2
667.8
693.0
717.4
724.5

Table 7.3: *Useful emission lines from mercury discharge lamps. Literature values for the lines measured* in vacuo *[1, 17, 18, 19] are compared with lines from a medium-pressure lamp measured with a Beckman Acta CV spectrometer [20].*

Literature wavelength measured in vacuo (nm)	Arc lamp wavelength (nm)
579.1	579.0
577.0	576.9
546.1	546.1
435.8	435.8
404.7	404.5
365.0	364.9
253.7	253.7

that these have usually been measured *in vacuo,* and the lines from a normal lamp can vary by up to 0.1 nm from these values.

At shorter wavelengths, the mercury lines at 253.65 and 184.96 nm from a low-pressure lamp are particularly useful. At the other extreme, lithium lines at 670.8 and 610.4 nm and potassium lines at 770.0 and 766.5 nm are conveniently placed. Neon has been recommended for the visible region [1] and Table 7.2 sets out some of these lines. The most-used region, between 250 and 580 nm, can be covered by a medium-pressure mercury lamp. Table 7.3 lists those bands which have been found by experience to be most readily detectable, sharp and free from shoulders. In Table 7.3, values obtained in our Laboratory are compared with *in vacuo* literature values.

Apart from these special sources, the deuterium emission lines in the visible region at 486.00 and 656.10 nm [2] are particularly useful for a quick check, since most spectrometers have a deuterium lamp as a source.

7.3   Absorption standards

Holmium and didymium (a mixture of neodymium and praeso-dymium) oxide-containing glasses have been used extensively as absorption wavelength standards over the past three decades. However they have now been shown to be rather unreliable as instrument quality improves, for the manufacture of reproducible glasses is

Table 7.4: *Wavelengths of selected absorption maxima for holmium and didymium glass filters [21]. Note that considerable variations are found between different batches of glass.*

Wavelength of maximum (nm)

Holmium glass	Didymium glass
241.5 ± 0.2	573.0 ± 3.0
279.4 ± 0.3	586.0 ± 3.0
287.5 ± 0.4	685.0 ± 4.5
333.7 ± 0.6	
360.9 ± 0.8	
418.4 ± 1.1	
453.2 ± 1.4	
536.2 ± 2.3	
637.5 ± 3.8	

Table 7.5: *Wavelengths of the absorption maxima of holmium(III) ions in perchloric acid reported by various workers: I. Milazzo et al., Cary 17 [22]; II. Vinter, Beckman Acta CV [20]; III. Burgess, Cary 118 [23]; IV. Burgess, Perkin-Elmer 200 [24].*

Wavelength of maximum (nm)

I [a]	II [b]	III [b]	IV [b]
241.15	241.0	241.1	241.1
249.75	250.0	249.7	249.7
278.2	277.8	278.7	278.2
287.15	287.5	287.1	287.2
333.5	333.3	333.4	333.3
345.6	345.5	345.5	345.0
361.5	361.0	361.5	361.2
385.6	385.6	385.5	385.6 [c]
416.2	416.0	416.3	416.6 [c]
450.7	450.4	450.8	451.0 [c]
452.0			
467.75			468.0
485.25	485.2	485.8	485.2 [c]
536.3			536.8
640.5			

a. The solution was 4 g of holmium oxide in 100 g 1.4M perchloric acid; *b.* The solution was 10 g of holmium oxide in 100 ml 17.5 per cent w/v aqueous perchloric acid. *c.* These are complex bands.

extremely difficult and, as a result, they are becoming difficult
to acquire. Their absorption bands, especially those of didymium,
are relatively broad and their positions are sensitive to slitwidth. The
NBS supply didymium glass filters as a wavelength standard, and
Venable and Eckerle [3] list 15 absorption maxima ranging from
402 to 784 nm that are used as calibration wavelengths. Procedures
for the location of the maxima are given and the effects of SSW and
temperature upon the apparent λ_{max} values described in detail.
Didymium glass filters have some useful bands in the near-UV but
usually become opaque at shorter wavelengths. Selected maxima
for typical holmium and didymium glass filters are given in Table 7.4:
other values can be found in [3-5]. The filters can only be
recommended for wide-band instruments (SSW > 5 nm) and the
uncertainties over the band positions should always be borne in mind.

An aqueous solution of rare-earth ions has narrower absorption
bands than the glasses [6]: an absorption spectrum of holmium
oxide in perchloric acid is given in Fig. 10.2. Tables 7.5 and 7.6

Table 7.6: *Wavelengths of the absorption maxima of
samarium(III) ions in perchloric acid reported by
various workers: I. Vinter, Beckman Acta CV [20];
II. Burgess, Cary 118 [23]; III. Burgess, Cary 219 [24];
IV. Burgess, Perkin-Elmer 200 [24].*

The solution was 10 g samarium(III) oxide in 100 ml
17.5 per cent w/v aqueous perchloric acid.

Wavelength of maximum (nm)

I	II	III	IV
305.4	304.8	305.2	304.8
317.5	316.6	317.2	317.1
332.2	331.2	331.9	331.6
344.8	344.2	344.2	344.1
362.7	361.9	362.2	362.1
374.5	374.1	374.0	374.1
391.2	389.8	390.4	390.1*
	400.9	401.0	401.2
407.5	406.4	406.6	406.7*
415.8	414.6	415.0	415.0*
417.7	416.6	417.0	416.8*
442.3	441.8	441.0	441.0
464.2	463.1	463.1	463.0*
479.6	478.4	478.5	478.9*
500.2	498.5	499.0	498.9*

* Multiple bands

set out the principal lines of holmium(III) and samarium(III) ions measured in various laboratories. Within the temperature range 15-35°C, both glasses and solutions of the rare-earth ions show little temperature variation, and this will only affect the second decimal place of the wavelength measured in nanometres [2]. A more important consideration is the SSW. A detailed study of some samarium and holmium lines as a function of bandwidth is presented in Table 7.7. Generally, an SSW that greatly exceeds the natural bandwidth of the line being measured will lead to a wavelength shift from the true value. Therefore the proper use of rare-earth ion solutions, or indeed emission lines from discharge lamps, is restricted to narrow-band instruments, i.e. of SSW less than 5 nm. The NBS has recently announced that they will be supplying ampoules of holmium oxide solution in perchloric acid as SRM 2034 with a calibration certificate giving λ_{max} values for 9 bands between 279 and 637 nm.

For far-UV calibration, oxygen and iodine vapour have absorption lines that are well characterized [7]. Iodine solutions have absorption maxima at 174.2, 174.8, 175.5 and 176.2 nm [8] but these solutions must be ultra-clean and free from oxygen [9].

The methods given in this and the preceding section are recommended as good wavelength standards. In the next section, more specialized procedures are outlined, and the excellent review article by Alman and Billmeyer [10] will be found to be a useful piece of supplementary reading.

7.4   Other methods

7.4.1   *The Van den Akker method for wide-band instruments*

For instruments of spectral slitwidth greater than 10 nm and for colorimeters, Van den Akker suggested a technique using a carefully chosen glass filter [11]. A mercury vapour emission line in the appropriate spectral region is selected and a filter is chosen whose transmission curve is approximately linear in this region, i.e. the transmittance over the bandpass λ_1 to λ_2 is a linear function of wavelength. This will not be true near peaks or troughs in the transmission curve. If the apparent transmittance of the filter is measured with an SSW of $\lambda_1 - \lambda_2$, then there will be a wavelength λ_c at which the same transmittance value would be seen if the measurement was done with a truly monochromatic source; λ_c will lie between λ_1 and λ_2 but will not necessarily be the mid-point.

Table 7.7: *The variation of the apparent* λ_{max} *values with slitwidth for narrow-band spectra. Solutions of holmium(III) oxide and samarium(III) oxide in perchloric acid measured on a Beckman Acta CV (linear wavenumber) instrument using various programmed slit settings [20].*

		Holmium bands			Samarium bands						
Nominal λ_{max} (nm)		361.0	245.5	333.3	401.0	374.5	362.7	344.8	332.2	317.5	305.4
NBW of band (nm)		4.2	3.6	3.3	—	4.2	3.9	4.1	3.3	4.1	3.7
Slit programme:											
I	SSW	0.1	0.1	0.1	—	0.1	0.1	0.2	0.2	0.2	0.2
	λ	361.3	345.7	333.6	—	374.9	362.65	344.8	332.2	317.7	305.8
II	SSW	0.1	0.2	0.2	0.3	0.3	0.35	0.45	0.6	0.7	0.7
	λ	361.3	345.4	333.4	401.1	374.2	362.3	344.5	331.95	317.5	305.4
III	SSW	0.3	0.35	0.4	0.5	0.6	0.8	1.1	1.5	1.6	1.5
	λ	361.3	345.4	333.5	401.2	374.7	362.7	344.6	332.0	317.7	305.5
IV	SSW	0.5	0.7	0.9	—	1.2	1.4	1.7	2.3	2.7	2.5
	λ	361.3	345.4	333.5	—	374.5	362.7	344.8	333.9	317.5	305.4
V	SSW	0.8	1.0	1.3	1.3	1.7	2.0	2.9	3.5	4.1	3.9
	λ	361.3	345.4	333.4	401.2	374.25	362.3	344.3	331.95	317.2	305.2
VI	SSW	1.0	1.4	1.7	2.7	3.5	4.1	5.8	7.0	8.2	8.2
	λ	361.3	345.3	333.3	401.2	373.8	361.8	343.6	331.7	316.45	303.95
VII	SSW	1.4	1.7	2.2	—	—	—	—	—	—	—
	λ	361.3	345.3	333.5	—	—	—	—	—	—	—
VIII	SSW	2.0	2.6	2.3	—	—	—	—	—	—	—
	λ	361.2	345.4	333.6	—	—	—	—	—	—	—
IX	SSW	4.1	5.2	6.4	—	—	—	—	—	—	—
	λ	361.3	345.4	333.3	—	—	—	—	—	—	—
X	SSW	8.1	8.1	8.1	—	—	—	—	—	—	—
	λ	361.0	345.1	329.8	—	—	—	—	—	—	—

NBW: Natural bandwidth of band in nanometres;   SSW: Spectral slitwidth in nanometres.

The filter is measured using the mercury source in place of the normal lamp. The normal lamp is then placed in position and the wavelength scale scanned until the same transmission value is observed: the wavelength scale then gives the value for λ_c. Van den Akker found that this value of λc should fall within 0.8 nm of the true wavelength of the line source. The method overcomes slit parameter errors and is dependent for its accuracy mainly upon the slope of the filter transmission at the calibration point. A slope of 1 per cent per nanometre represents a 0.1 nm uncertainty in the final value of λ_c.

7.4.2 *Interference filters*

Heidt and Bosley [12] have described the manufacture and use of an interference cell that consists of a pair of closely-spaced parallel plates that generates a transmission spectrum that is a series of maxima and minima. The cell is calibrated by using a line source in the same instrument. Any part of the spectrum can be covered by varying the separation of the plates. Since the interference maxima are rather broad, it is difficult to locate the peak exactly. The slit function must be symmetrical, and a temperature effect will be seen if the cell spacer is liable to expand.

Buist has pointed out that an ordinary infrared cell equipped with fused silica or, better, calcium fluoride windows will give these fringes, and with a 25 μm spacer it is possible to generate fringes over the 220-850 nm range [13].

7.4.3 *Retardation plates*

By constructing a cell consisting of a Nicol prism polarizer, a quartz birefringent plate and a Nicol prism analyser, Buc and Stearns were able to calibrate their spectrometer in a non-empirical way without recourse to calibration materials [14]. The method depends upon the fact that the quartz plate splits plane-polarized light into two orthogonal plane-polarized beams of differing velocity. Depending upon their phase relationship at the exit surface, a series of transmission maxima and minima are seen across a large portion of the UV-visible range. The positions of these maxima can be calculated from a knowledge of the cell dimensions and the birefringence of quartz. The temperature must also be known and, although the slit function should be symmetrical, its width and slope are irrelevant. The authors were also able to use this method to estimate slit shapes and widths.

7.4.4 *The cross-filter technique*

This method involves the use of two filters whose transmission curves cross at some point [15]. This point is determined on an instrument that has already been calibrated. The method is susceptible to slit variation, it is a one-point calibration and it would be necessary to determine whether the transmission curve of either filter changed with temperature.

7.5 Conclusion

For narrow-band instruments, that is, of SSW less than 5 nm, line sources are the most effective means of wavelength calibration. When making the calibration, as Robertson [16] has pointed out, peripheral equipment should also be checked for performance, e.g. the response speed of pen recorders, for this can introduce errors into the calibration. For wide-band instruments, the rare-earth glasses are convenient to use. Van den Akker's procedure is of particular relevance when studying materials with very broad absorption bands, such as those encountered in colorimetery. The interference and retardation techniques are probably best left to specialists.

Calibrations using a mercury arc lamp and holmium perchlorate solutions are described in detail in Chapter 10.

References

1 Menzies, A.C. (1960), *Pure & Appl. Chem.*, **1**, 147.
2 Reule, A.G. (1976), *J, Res. Nat. Bur. Std.*, **80A**, 609.
3 Venable, W.H. and Eckerle, K.L. (1979), NBS Special Publication No. 260-66.
4 McNeirney, J. and Slavin, W. (1962), *Appl. Opt.*, **1**, 365.
5 *UV Atlas of Organic Compounds* Perkampus, H., Sandeman, I. and Timmons, C.J. (Eds) (1971), Verlag Chemie.
6 West, M.A. and Kemp, D.R. (1976), *Int. Lab.*, May/June, 27.
7 Kaye, W.I. (1961), *Appl. Spec.* **15**, 89.
8 Cordes, H. (1935), *Z. Physik*, **97**, 603.
9 Bramston-Cook, R. and Erickson, J.O. (1973), Varian Instruments Applications, **7**, No. 3.
10 Alman, D.H. and Billmeyer, F.W. (1975), *J. Chem. Educ.*, **52**, A315.
11 Van den Akker, J. (1943), *J. Opt. Soc. Amer.*, **33**, 257.
12 Heidt, L. J. and Bosley, D.E. (1953), *J. Opt. Soc. Amer.*, **43**, 760.
13 Buist, G.J. (1976), *J. Chem. Ed.*, **53**, 727.

14   Buc, G.L. and Stearns, E.I. (1945), *J. Opt. Soc. Amer.*, **35**, 465.

15   Sanghi, I. and Parthasarathy, N.V. (1959), *Naturwiss.*, **46**, 315.

16   Robertson, A.R. (1976), *J. Res. Nat. Bur. Std.*, **80A**, 625.

17   Harrison, G.R. (1969), 'Wavelength Tables.' MIT Press, Cambridge, Mass.

18   Zaidel, A.N. (1961), *Tables of Spectrum lines*, Pergamon, p. 381.

19   Gibson, K.S. (1949), NBS Circular No. 484, pp. 13-15.

20   Vinter, E., unpublished observations.

21   Edisbury, J.R. (1966), *Practical Hints on Absorption Spectrometry*, Hilger & Watts, London.

22   Milazzo, G. (1976), *Subcommittee on Calibration and Test Materials*, IUPAC commission on physicochemical measurements and standards.

23   Burgess, C. (1977), *UV Spec. Grp. Bull.* **5**, 77.

24   Burgess, C., unpublished observations.

8 Cell handling

8.1 Handling

Care should be taken to preserve the optical surfaces of the cell. Dry cells should be handled with clean cotton gloves, avoiding touching the windows. Wet cells should be handled by plastic or padded tweezers. Particular care should be taken not to insert into a cell any object which could scratch the optical surfaces.

For measurements of the highest quality, the windows of the cell should never be touched; if the windows become contaminated in use, the cell should be re-cleaned using one of the procedures given below. Wiping can damage the windows and reduce their UV-transmission. However, if windows are wiped, this should be done with several thicknesses of a good quality lens tissue - not normal cleaning tissues. The tissue should not be pressed on the window with the fingers, as grease from the fingers can penetrate the tissue and cause worse contamination. Normal cleaning tissues can contain abrasive particles and fluorescent brightening agents which absorb strongly in the UV region. If the transmission of a cell after cleaning is less than expected, it is prudent to check the performance of the instrument before re-cleaning the cell.

8.2 Filling and emptying

A cell should be filled and emptied while in its cell-holder. Cells should be filled using transfer pipettes, which should be designed so that it is impossible to overfill the transfer pipette or the cell; a suitable design is shown in Fig. 8.1. It is good practice to take as much care over the cleanliness of the transfer pipette as that of the cell.

Tests show that most cells can be emptied more efficiently by

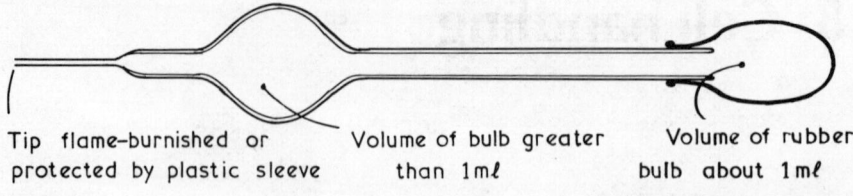

Tip flame-burnished or   Volume of bulb greater   Volume of rubber
protected by plastic sleeve   than 1 mℓ   bulb about 1 mℓ

Fig. 8.1: *A specially-made glass transfer pipette for filling 10 mm normal rectangular cells.*

sucking out the sample than by inverting the cell. A thin plastic tube is inserted into the cell and the sample withdrawn by gentle suction. This procedure is carried out as quickly as possible to limit evaporation of the liquid film on the windows.

8.3   Cleaning

The mildest method giving satisfactory results should be used and, if drastic methods are necessary, they should be carried out quickly since prolonged contact with strong agents may damage the windows. Cells should be rinsed immediately after use, especially when measuring biological samples, colloids and other materials which tend to coat the cell windows.

8.3.1   *Plastic cells*

Users requiring to clean plastic cells should refer to the manufacturers of the cell. Cold solutions of mild detergents are generally most suitable, and organic solvents, oxidizing acids or hot cleaning agents should not be used.

8.3.2   *Cemented cells*

Cold or hot solutions of mild detergents are generally most suitable. The manufacturers should be consulted if organic solvents or oxidizing agents are to be used.

8.3.3   *Fused glass or silica cells*

The methods given below are arranged in order of 'strength'. Alkaline solutions will damage both glass and silica windows and should be avoided if possible.

(a) Rinse with water or organic solvent. This is often satisfactory between measurements, but over longer periods there may be a build-up of contamination. Many organic solvents contain aromatic compounds with high UV absorption and appreciable fluorescence.

(b) Cold or hot solutions of mild detergents. Check that the detergent is not alkaline in nature.

(c) Cold fuming nitric acid. This is a strong oxidizing agent that is effective in the cold.

(d) Hot nitric acid. An oxidizing agent that is generally as effective as chromic acid.

(e) Ultrasonic agitation in nitric acid. This is effective but should be applied with caution as cells may be shattered by ultrasound.

(f) Hot trisodium orthophosphate. This is alkaline and will etch windows if the treatment is prolonged or used frequently.

If all else fails, carry out the following sequence of operations:

1. Boil or agitate in concentrated nitric acid, rinse thoroughly with distilled water.

2. Immerse in hot (80°-90°C) trisodium orthophosphate for 10 minutes, rinse in several changes of hot distilled water.

3. Immerse for 10 minutes in boiling distilled water, followed by rinsing with cool ethanol.

4. Rinse thoroughly in hot, freshly distilled ethanol.

CAUTION: All of these cleaning agents should be treated with respect, but fuming nitric acid and hot nitric acid are powerful oxidizing agents and must be handled with extreme caution. Goggles, gloves and other protective clothing should be worn, and operations carried out in an efficient fume cupboard. Containers should be plainly marked. Nitric acid reacts violently with many organic compounds, particularly alcohols. Cells containing organic material should be rinsed with an organic solvent miscible with water, such as acetone or alcohol, and then thoroughly rinsed with water before putting them into nitric acid.

8.4 The drying of cells

If cells are in regular use with aqueous solutions it is best not to attempt to dry them internally. If this is necessary, the cell should first be rinsed thoroughly with water and then several times with spectroscopic-grade methanol (commercial ethanol often contains water or aromatic hydrocarbons). Cells should not be touched with the fingers during this operation. The cell should then be left upside-down to drain, and will dry fairly quickly. A jet of clean air, free of dust and oil, or of nitrogen will promote drying.

8.5   Storage of cells

Cells should be stored in a clean condition and in such a way that they remain clean and are protected against damage. The velvet-lined boxes supplied by some manufacturers harbour dust and touch the optical faces of the cell, and so are not recommended. In general, it is best not to dry cells but to store them in a wet state after cleaning. Thus they should be stored immersed in water or an appropriate organic solvent in cylindrical jars or tubes, so that the windows cannot be damaged. It is essential that flow and sampling cells are kept filled at all times.

Cells should be dried and wrapped in lens tissue for long-term storage. They may require cleaning before use.

Cells should never be stored with the stoppers in place. Temperature changes cause differential expansion of different materials and lead to seizing of the stoppers or even the cracking of the cell.

8.6   Cell matching

Most manufacturers will supply cells in matched sets of two or more cells based on the requirement of BS 3875, namely, with transmittance differing by not more than 1.5 per cent at 240 nm for fused silica cells, and 0.5 per cent at 365 nm for glass cells. The knowledge that cells are matched is an aid to obtaining flat instrumental baselines when the cells are used for sample and reference, or for the quick comparison of different samples when placed in matched cells. However, the cells will only show the degree of matching if they are scrupulously cleaned, and the fact that the cells are matched in no way avoids the necessity of comparing them when filled with solvent each time they are used.

In general, the uniformity of synthetic fused silica and its low absorbance down to 180 nm means that cells of this material fall within the BS 3875 limits without special selection. The transmission of different batches of glass and fused quartz show sufficient variation to necessitate careful selection of paired cells.

8.7   Substitution methods

When measurements are made with a double-beam instrument, the greatest accuracy can be attained by keeping the cell fixed rigidly in the cell holder and rinsing and filling the cell with successive solutions without moving it. The cell should be periodically filled with pure solvent to check for instrumental drift or the build-up of contaminants. Care must be taken in filling and emptying the cell to ensure that the solution does not run down the outside of the cell.

8.8   Sampling cells

These should be filled and emptied according to the manufacturer's instructions. If this is done, the amount of sample left in the cell after emptying should be less than 1 per cent of the working volume of the cell. If highly absorbing solutions are being measured, the cell should be filled with solvent or the new solution several times and emptied before being filled for the next measurement.

8.9   Flow cells

The scavenging characteristics are determined by many factors: cell design, flow rate, viscosity of solvent, etc. Under ideal conditions, a sample should be completely displaced by a second solution when 3 cell-volumes of the latter have passed through. However, in practice, the user should check the performance of the system by filling the cell with a highly absorbing sample and checking the rate at which it is displaced when pure solvent is passed through under normal operating conditions.

9 Tests of cell performance

9.1 Measurement of pathlength

9.1.1 *Mechanical methods*

Open-topped cells of pathlength greater than 5 mm can be measured with vernier calipers, though great care must be taken to avoid scratching the windows. The pathlength at the top of smaller cells can be measured with a taper gauge, but this can only be related to the pathlength at the working area by the assumption that the inside faces of the windows are parallel.

If a large number of cells of a certain pathlength are to be measured, an air gauge can be usefully employed. The probe of the gauge is a block of width slightly less than the pathlength to be measured. Air is forced through holes in the opposite faces of the blocks, and the back-pressure is related to the size of the gaps between the block and the cell windows. The air cushion reduces scratching of the windows.

The National Bureau of Standards have an 'electronic' feeler gauge with a long probe. This is mounted on a micrometer test-bed and is used to measure the inside dimensions of stoppered Normal cells. The set-up is illustrated by Mavrodineanu and Lazar [1].

Measurement of the pathlength of a 10 mm cell can be made with a resolution of 0.125 μm. In general, stoppered or cylindrical cells can only be measured from the outside. A travelling microscope is probably the best way of doing this. It is often difficult to see the exact position of the inside face of the window when viewed from outside. This is made easier if the empty cell is immersed in a liquid of high refractive index, e.g. carbon tetrachloride. Alternatively, two ball bearings of known diameter are put inside the cell. They are held together against one face by means of a magnet applied to the outside (Fig. 9.1). Measurement is made from the point of contact

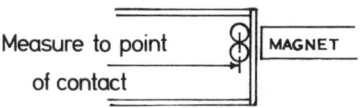

Fig. 9.1   *The use of ball-bearings and a magnet in the measurement of pathlength.*

of the balls. They are then moved to the opposite window to give its location. The diameter of the balls is added to the difference between the two measurements to give the pathlength.

9.1.2   *Photometric methods*

Theoretically an absolute determination of cell length could be performed by filling the cell with a solution of known concentration of a solute of known extinction coefficient, and measuring its absorbance: suitable solutions are given in Chapter 4. However it is unlikely that an accuracy of better than 2 per cent could be attained by this method, and a comparison technique is to be preferred. In this, the absorbance of the test cell filled with a suitable solution is compared with the absorbance of a Grade A cell, or an open-topped cell that can be measured by mechanical means. The solution should be chosen to give an absorbance reading in the optimum range of the instrument - usually 0.5 to $1A$ for modern instruments - and the two cells should be of similar pathlength so that the same part of the absorbance scale is used for both.

9.1.3   *Optical methods*

Interferometric methods can be used, though the equipment and expertise are not commonly available. Very short pathlength cells (up to 1 mm) can be measured by means of infrared interference fringes generated in a conventional IR spectrometer. The technique is described by Buist [2].

9.2   **Measurement of transmission**

There is an appreciable difference between the apparent transmission of cells measured empty and filled with a liquid. This is primarily due to a reduction in the reflection at the inner surfaces of the windows when the cell is filled with a liquid of refractive index greater than air, but scattering and absorption by the liquid will now have an effect, particularly at shorter wavelengths.

Since the cells described in this Volume are normally used filled with liquid, and since the most commonly-used solvent is water, it is recommended that transmission measurements be made with the cell filled with water. The water should be freshly distilled and free from dissolved gases, absorbing ions and organic matter. The UV absorbance of freshly-distilled water increases on standing in air, possibly due to the absorption of carbon dioxide, and so it should be used immediately. Samples of distilled water prepared and measured under normal laboratory conditions have been found to show some absorption in the far UV and above 800 nm. A cell filled with water may, therefore, have a lower transmission below 200 nm and above 800 nm than when measured empty.

The cell should be scrupulously cleaned inside and out, rinsed with the freshly-distilled water and then completely filled. It should be fitted with a lid or stopper if possible. The 100 per cent transmission line of the instrument should be measured with empty cell-holders in reference and sample positions, and the test cell then placed in the sample position in the correct orientation.

The measurement should be carried out as quickly as possible and, with reference to the transmission curves of Chapter 2, particular care taken in the regions of rapid change near the cut-off wavelength.

9.3   Intrinsic fluorescence of cell windows

Cells with windows containing metallic impurities show considerable fluorescence; for example, Fig. 2.8 shows the prominent 395 nm emission band from a fused quartz cell excited with 250 nm radiation. A measurement of the intrinsic fluorescence of a cell should be made with the cell filled with freshly-distilled water. This has the advantage that, if an instrument of good resolving power is used, the water Raman band will be seen and can be used as a reference of intensity in assessing the window emission.

The emission comes primarily from the entrance window of the cell and the amount of this emission reaching the detector will depend upon the design of the instrument and the operating parameters. Thus a calibration applies only to the cell in a particular orientation in a particular instrument under specific operating conditions.

9.4   Angular deviation and dispersion

Angular deviation is caused by non-parallelism of the windows or the two faces of a particular window. This can be readily checked as

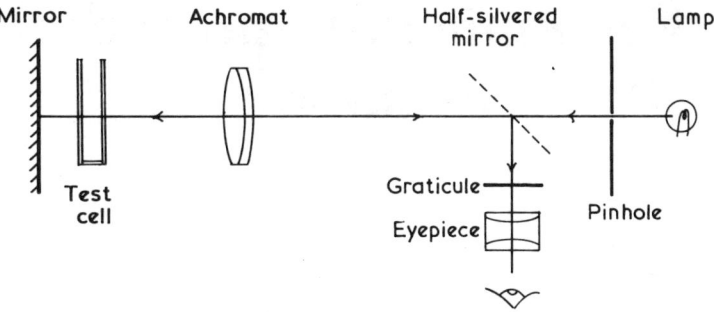

Fig. 9.2   *Diagram of an apparatus for the assessment of dispersion and deviation in a spectrophotometer cell.*

follows: project the image of a pin-hole on a screen; place the cell in front of the projector lens at a distance of 3.44 metres from the screen. The cell should be masked so that light passes only through the working area. Measure the displacement of the spot - a displacement of 1 mm corresponds to 1' of arc.

Dispersion, caused by refractive errors in the cell windows, is more difficult to measure and to specify. A leading manufacturer uses the apparatus shown in Fig. 9.2 to measure both deviation and dispersion. Again, light from a pin-hole is projected through the cell and is then reflected back again to a graticule that has a series of concentric rings. The displacement of the centre of the image gives the deviation, while the distortion - an increase in the effective diameter - of the image is a measure of the dispersion. In routine use, the operator determines whether any of the image is outside a particular circle on the graticule.

References

1   Mavrodineanu, R. and Lazar, J.W. (1973), *Clin. Chem.,* **19**, 1053.
2   Buist, G.J. (1976), *J. Chem. Educ.,* **53**, 727.

10 Recommended procedures for standardization

10.1   Resolution of monochromators

The best performance of a spectrometer will only be attained - in terms of both absorbance and wavelength accuracy - if careful consideration is given to the resolution of the monochromator. Since resolution is a function of slitwidth as well as dispersion of the instrument, the choice of slit setting is a critical one. The smaller the slitwidth, the greater the resolution, but the corresponding reduction in energy means that the signal-to-noise ratio falls. It is therefore necessary to select the smallest possible slitwidth that gives an acceptable noise level.

When measuring an absorbance band in a high-resolution instrument, it is recommended that the spectral slitwidth (SSW) should not exceed 10 per cent of NBW of the band (see Fig. 1.1). the slit setting is given by SBW/dispersion. The dispersion of a grating instrument is the same at all wavelengths, while that for a prism instrument must be read from the manufacturer's data at the appropriate wavelength.

There are two simple checks for the resolution of an instrument:

(a) The benzene vapour absorption spectrum has a minor peak at 259.6 nm that can be resolved from the main band if the instrument has an SSW of less than 0.5 nm (Fig. 10.1). A drop of benzene is placed in a 10 mm stoppered cell and allowed to come to equilibrium at room temperature. The vapour pressure is such that the absorbance at the test wavelength will probably be about $0.8A$.

(b) The profile of the 656.1 nm line from the emission of the instrument's deuterium lamp is measured using the single-beam or 'energy' mode of the instrument. The apparent width of the band at half-peak height is taken to be the SSW of the instrument.

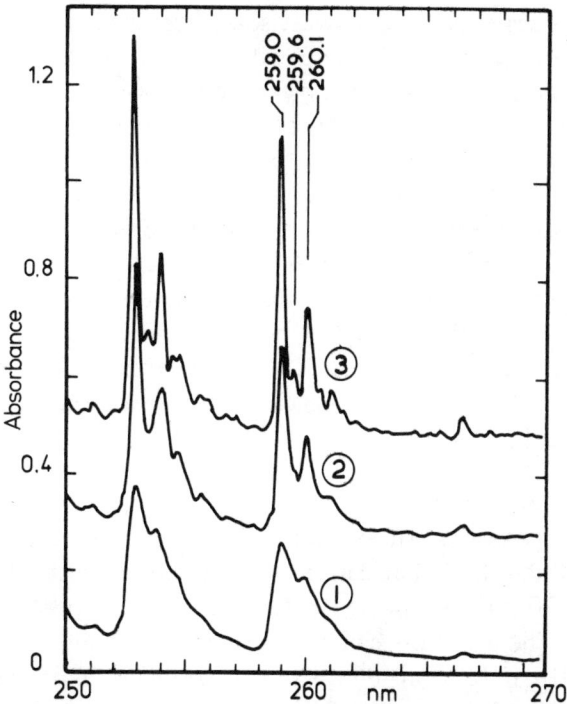

Fig. 10.1   *A portion of the benzene vapour absorption spectrum measured at three different spectral slitwidths in a Perkin-Elmer 402 at a scanning speed of 10 nm min* $^{-1}$ *:1. SSW = 2.0 nm; 2. SSW = 1.0 nm; 3. SSW = 0.5 nm. Numerical values taken from Lang [1]. The curves are displaced vertically for clarity.*

The SSW of the instrument represents the minimum NBW of a compound that can be measured at that wavelength with any degree of confidence.

10.2   Wavelength calibration

10.2.1 *High-accuracy calibration*

A discharge lamp is recommended for this purpose. A low-pressure mercury lamp has a number of intense lines that cover a large part of the UV and visible range. The lamp should be placed as near as possible to the entry slit of the monochromator. 'Pen-ray' discharge lamps are well suited to this purpose.

The instrument should be operated in single-beam or 'energy' mode, with slitwidths as small as possible. With recording instruments, care should be taken to relate the chart divisions accurately to the wavelength scale and the slowest scanning speed should be used. All measurements should be made with the instrument scanning in the normal direction to avoid the effects of back-lash. Since temperature affects the wavelength setting of most instruments, the instrument should be allowed to warm up to its normal operating temperature, and the ambient temperature noted.

The most useful lines are noted in Table 7.3, namely 579.0; 576.9; 546.1; 435.8; 404.5; 364.9 and 253.7 nm. A correction chart can be constructed by plotting the differences between the observed positions of the lines and the positions given above against wavelength. For grating instruments, these differences should not exceed ±1 nm. For prism instruments, they should not exceed ±1 nm below 400 nm and ±2 nm above.

10.2.2 *Routine calibration*

The peak positions of a solution of holmium(III) ions form a convenient standard that can be read to the nearest 1 nm with ease. Dissolve 0.50 g of holmium oxide in 2.4 ml AR perchloric acid (72 per cent). The solution can be warmed to aid dissolution, but it is better to stir it at room temperature overnight. This solution is then diluted to 10 ml with water. Scan the spectrum at low speed with the minimum slit setting. A typical spectrum is given in Fig. 10.2. Locate the following eleven prominent lines, given to the nearest 0.5 nm: 241.0; 250.0; 278.5; 287.0; 333.0; 345.5; 361.5; 385.5; 416.5; 451.0; 486.0 nm.

10.3   Stray-light measurement

A complete analysis of the stray-light characteristics of an instrument requires the measurement of monochromator stray-light, but since this is a tedious process requiring specialized equipment, it is not usually undertaken. Instrumental stray-light (ISL) is more significant as far as stray-light errors are concerned, and depends not only on momochromator stray-light but the absorbance of the sample being measured and the response of the detector.

10.3.1 *Routine measurement of ISL*

The cut-off filter is satisfactory for the majority of routine applications. It must always be borne in mind that the ISL is a function of the sample: the measurement of x per cent ISL with a cut-off filter does not mean that x per cent will again be present

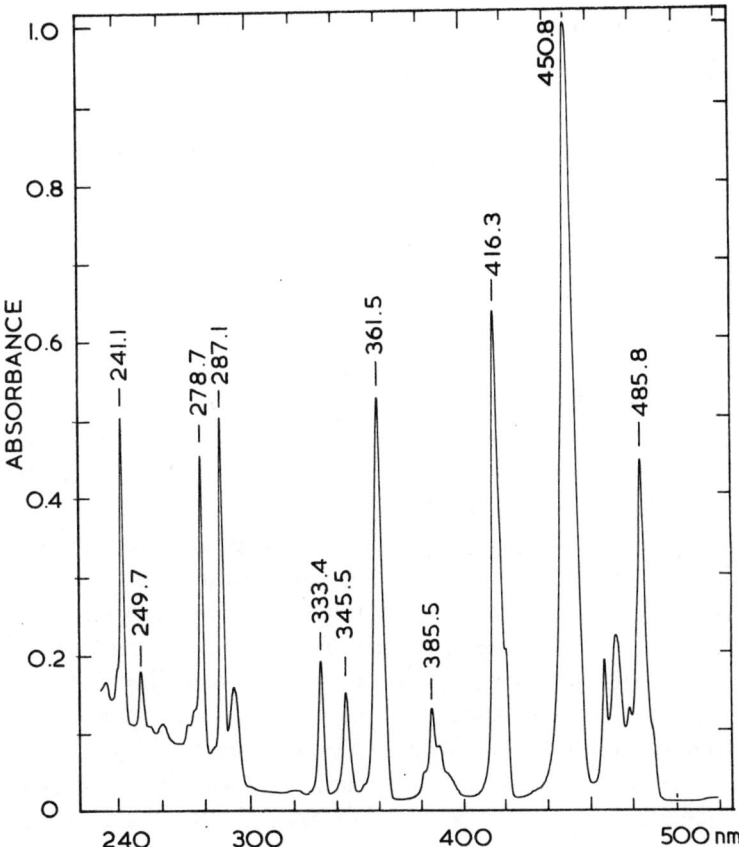

Fig. 10.2 *Absorption spectrum of a 5 per cent solution of holmium (III) perchlorate in 17.5 per cent perchloric acid with 10 mm pathlength. Recorded with a Perkin–Elmer–Hitachi 200 at a scanning speed of 60 nm min^{-1}, SSW = 1.0 nm [5].*

when a different absorber is in the beam (see Section 6.4.2). It is better to regard the filter method as one which detects stray-light rather than measures it.

The solutions and liquids listed in Table 10.1 are recommended as standard cut-off filters. They are also the recommendations of the American Society for Testing and Materials (ASTM) [2], and are generally accepted as industrial standards. Compared with glass filters they have the advantages of reproducibility and freedom from fluorescence. It should be noted that the cells used must be clean, free from fluorescence and with as high a transmission as possible in the region under investigation. Attention to these factors is particularly important when measuring stray-light below 220 nm.

Table 10.1: *Cut-off filters for stray-light tests (data from [2]).*

Spectral range (nm)	Liquid or solution	Pathlength (nm)
165-173.5	Water	'0.10
170-183.5	Water	10.0
175-200	Aqueous KCl (12 g l^{-1})	10.0
195-223	Aqueous NaBr (10 g l^{-1})	10.0
210-259	Aqueous NaI (10 g l^{-1})	10.0
250-320	Acetone	10.0
300-385	Aqueous $NaNO_2$ (50 g l^{-1})	10.0

The absorbance of the first four of these filters is strongly affected by dissolved oxygen. Pure nitrogen should be bubbled through for several minutes before use and the water should be freshly distilled. Water purified by ion-exchange methods may contain significant amounts of organic impurities. Different concentrations or path-lengths may be used to displace the absorption edge of these filters so that they can be used in other regions. The absorbances of the solutions increase with temperature by about 2 per cent per °C [3]. The NBS now supplies crystalline potassium iodide (SRM 2032) as a standard for the assessment of stray-light. Absorptivity values are given at 5 nm intervals from 240 to 275 nm.

Saturated solutions of lithium carbonate are commonly used for stray-light tests in the range 210-225 nm [4], but the concentration of the saturated solution is temperature-dependent, and so this solution is not recommended.

When measuring very low stray-light levels, it is necessary to attenuate the reference beam with a metal screen in order to extend the absorbance range of the instrument. Details of this procedure are given by the ASTM [2]. For the measurement of stray-light above 700 nm, a filter is required whose cut-off region extends to longer wavelengths. There are no suitable liquid or solution filters, so a glass filter must be employed. Heat-absorbing glass is fairly satisfactory; a sample tested had a rather broad cut-off extending from 82 per cent transmittance at 600 nm to less than 0.5 per cent at 1000 nm. This is far from ideal; nevertheless the filter gave a useful indication of stray-light at 900 nm in two single-monochromator grating spectrophotometers, which proved to be 15-20 per cent in each case. To evaluate a filter for use above 700 nm, its true absorbance must be measured by a spectrometer equipped with a lead sulphide detector.

10.3.2 *Quick stray-light checks*

Vycor glass has been used extensively as a cut-off filter for stray-light measurement in the range 200-210 nm, but it is unsuitable as a standard because its absorption edge is not very steep, different samples may not have exactly the same absorption characteristics and there may be interference from the fluorescence of the glass. However it is useful as a check in this spectral range when the stray-light is relatively high, i.e. greater than 1 per cent.

10.4   Absorbance standards

10.4.1 *Photometric linearity*

Several linearity standards were mentioned in Chapter 4, for example, the green food dyes, and since these all obey Beer's Law, they may all be used to check instrument response by serial dilution. Alternatively, a single concentration can be used in a series of cells of increasing pathlength, say 5 mm to 40 mm. This procedure is recommended since the tolerance on cell pathlength is far smaller than the uncertainties associated with volumetric manipulations.

10.4.2 *Photometric accuracy*

(a)   *High accuracy calibration*

For the most exacting work, a set of neutral glass filters or Nichrome-on-fused-quartz filters calibrated by the NBS or NPL is recommended. The conditions for their use will be detailed by the calibration service.

(b)   *Routine calibration*

It is recommended that potassium dichromate in 0.01N sulphuric acid is used as the standard. AR potassium dichromate is dried at 110°C for 1 hour. Alternatively, this can be purchased as SRM 935 (see Section 4.5.5). Solutions in 0.01N sulphuric acid at two concentrations should be made:
Solution A: 50 mg ± 0.5 mg in 1 litre for the absorbance range 0.2-0.7A;
Solution B: 100 mg ± 1 mg in 1 litre for the range 0.4-1.4A.

Measurements should be made in 10 mm cells with the temperature controlled in the range 15-25°C, using 0.01N sulphuric acid as the reference. Table 10.2 gives the expected values for the two solutions at the two maxima and minima of the solutions based on literature values. The tolerances represent the range of acceptable values,

Table 10.2: *Recommended absorbance values for acidic potassium dichromate solutions.*

Wavelength (nm)	Absorbance	
	Solution A	Solution B
235 (min.)	0.626 ± 0.009	1.251 ± 0.019
257 (max.)	0.727 ± 0.007	1.454 ± 0.015
313 (min.)	0.244 ± 0.004	0.488 ± 0.007
350 (max.)	0.536 ± 0.005	1.071 ± 0.011

based on the uncertainties of the literature values and the temperature coefficient over this temperature range.

References

1 Lang, L. (1966), *Absorption Spectra in the UV and Visible Region,* Akademiai Kiado, Budapest.
2 ASTM Standard E-387, Annual Book of ASTM Standards, 1978.
3 Slavin, W. (1963), *Anal. Chem.,* **35**, 561.
4 Hartree, E.F. (1963), *Photoelec. Spectr. Grp. Bull.,* **15**, 398.
5 Burgess, C. (1977), *UV Spec Grp. Bull.,* **5**. 77.

Appendix Selected publications of the UV spectrometry group

A.1 Absorbance and wavelength standards

Knowles, A. (1978), Standards Working Party report meeting. *UVSGB*, **6**, 56.

Burgess, C. (1977), Monitoring the performance of UV-visible spectrophotometers. *UVSGB*, **5**, 77.

Douglass, S.A. and Emary, R.J. (1977), Standardization of temperature, absorbance and wavelength measurements in UV-VIS spectrophotometry. *UVSGB*, **5**, 85.

Popplewell, B.P. (1977), The calibration of neutral density filters. *UVSGB*, **5**, 90.

Knowles, A. (1977), A new commercial system of solutions for spectrophotometer checks. *UVSGB*, **5**, 94.

Clarke, F.J.J., Davis, M.J. and McGivern, W. (1977), Transmittance standards from NPL. *UVSGB*, **5**, 104.

Sharpe, M.R. (1975), Some investigations into the use of filters for calibrating UV-VIS spectrophotometers. *UVSGB*, **3**, 57.

Johnson, E.A. (1967), Potassium dichromate as an absorbance standard. *PSGB*, **17**, 505.

Everett, A.J., Young, P.A. *et al.* (1965), Official report on the PSG collaborative test of recording spectrophotometers. *PSGB*, **16**, 443.

Tarrant, A.W.S. (1965), Some comments on the findings of the PSG collaborative test. *PSGB*, **16**, 458.

Glenn, A.L. (1965), Further comments on collaborative tests. *PSGB*, **16**, 464.

Neal, W.T.L. (1956) Differential absorptiometry. *PSGB*, **9**, 204.

Ketelaar, J.A.A., Fahrenfort, J., Haas, C. and Brinkman, G.A. (1955), The accuracy and precision of photoelectric spectrophotometers. *PSGB*, **8**, 176.

Morton, R.A. (1951), Collabroative test on potassium dichromate - introductory remarks. *PSGB*, **4**, 65.

Gridgeman, N.T. (1951), Statistical analysis: the accuracy and precision of photoelectric spectrophotometry. *PSGB*, **4**, 67.

Harding, H.G.W. (1951), Instrumental aspects: precautions necessary for accurate measurements of O.D. standards. *PSGB*, **4**, 79.

Lothian, G.F. (1951), Notes on future spectrophotometric tests. *PSGB*, **4**, 86.

Edisbury, J.R. (1950), Further comments on the P.S.G. collaborative test with potassium nitrate. *PSGB*, **2**, 32.

Edisbury, J.R. (1949), Collaborative test: relative readings on twenty-eight Beckman spectrophotometers. *PSGB*, **1**, 10.

A.2 Stray-light

Knowles, A. (1978), Stray-light in grating monochromators. *UVSGB*, **6**, 84.

Sharpe, M.R. and Irish, D. (1976), Stray-light in grating spectrophotometers. *UVSGB*, **4**, 51.

Hartree, E.F. (1963), Stray-light in ultraviolet spectrophotometers: the need for a standard criterion. *PSGB*, **15**, 398.

Lothian, G.F. (1956), Effects of finite slitwidth and stray radiation in differential absorptiometry. *PSGB*, **9**, 207.

Cannon, C.G. (1955), Anomalous slitwidth effect in differential absorptiometry. *PSGB*, **8**, 201.

Anon. (1952), Proposed collaborative stray-light test. *PSGB*, **5**, 119.

Collins, F.D. (1951), Notes and observations: a postscript on stray-light. *PSGB*, **4**, 96.

Perry, J.W. (1950), Sources and treatment of stray-light in spectrophotometry. *PSGB*, **3**, 40.

Donaldson, R. (1950), Measurement of stray-light by double monochromator. *PSGB*, **3**, 45.

Martin, A.E. (1950), Stray-light in infrared spectrometry. *PSGB*, **3**, 50.

A.3 Cells

Knowles, A. (1978), Cell Working Party report meeting. *UVSGB*, **6**, 54.

Goddard, D.A. (1976), The cleaning of vitreous silica cells. *UVSGB*, **4**, 19.

Archer, M.S. (1954), Techniques for using absorption cells in ultraviolet spectrophotometry. *PSGB*, **7**, 160.

Abbreviations: *UVSGB: UV Spectrometry Group Bulletin; PSGB: Photoelectric Spectrometry Group Bulletin.*

Index